GEOLOGY

Field Guide to Southern California

Robert P. Sharp
California Institute of Technology

GEOLOGY
Field Guide to Southern California

WM. C. BROWN COMPANY PUBLISHERS
Dubuque, Iowa

THE REGIONAL GEOLOGY SERIES

Consulting Editor

John W. Harbaugh
Stanford University

CONTENTS

FOREWORD

Dr. Sharp's book is one of a series of books in the Wm. C. Brown *Regional Geology Series*. The objective of this series is to provide an authoritatively written layman's guide to important geologic features in each region treated in the series. Stress is placed on observations in the field. Each guide provides an overview of the geologic provinces of the region to which it pertains, and outlines a series of self-guiding field trips which will allow users to make their own, first-hand observations on features that typify the provinces.

The series is directed toward diverse groups of users. The series should find use in formal classes in geology, both at the college and university level, and in high schools, in which field trips form an essential part of introductory or advanced courses. Furthermore, books in the series should be useful to professional geologists and other scientists who desire an introduction to the geologic features of particular regions. Finally, the series should find use among individuals who are not necessarily trained in science, but who do have an active interest in natural history and who enjoy travel.

Authors of books in this series all have intimate acquaintance with their respective regions and extensive teaching experience which has stressed field-trip observations. Consequently, each book in this series represents a distillation of teaching experiences that have involved many students and numerous field trips.

John W. Harbaugh
Consulting Editor

PREFACE

Most things are interesting, even old bleached cow bones, provided you know something about them. Indeed, much of the pleasure and enjoyment of life come from knowledge, familiarity, and understanding of things, be they related to sports, music, drama, science, art, birds and bees, or flowers and trees. Among scientists, geologists are reputed to have as much fun as anybody because of their understanding and appreciation of the natural environment. An objective of this effort is to share that fun with you.

Consisting of three parts, this booklet is designed for people without any formal acquaintance with geology but with an inherent interest in nature and the out-of-doors. The first chapter aims to provide some understanding of basic geological matters; it can well be skipped by those with previous knowledge of the subject. The second furnishes descriptions of geological features and relationships in nine natural provinces within southern California. The third, constituting more than half the booklet, is a guide to geological features visible along two travel routes, one from Los Angeles Basin to Death Valley and the other from Los Angeles Basin to Mammoth up the east side of the Sierra Nevada. Emphasis is upon geology that can be seen while traveling by car along highways, often at a fair rate of speed.

In spite of current interest in the moon and other planets, most of us are going to spend all of our life on the surface of planet Earth, and we will enjoy our stay more if we know something about aspects of the natural environment. A speaking acquaintance with the surrounding rocks, geological structures, and landforms can be a source of real satisfaction and enjoyment. We aim to excite your interest in these matters and to impart some information about them at the same time.

Use of technical jargon is minimal but avoidance of all geological terms is impossible. A foreign language cannot be read without at least some vocabulary; so a glossary is provided for necessary terms (Appendix B). A booklet of this size cannot possibly cover all the geology of southern California, so a high degree of selection has necessarily been exercised. This is not an "all about" book, but hopefully you will find it a fun book. Most of you already know more geology than you realize.

I am deeply indebted to Dorothy L. Coy for skillful and patient typing and retyping of manuscript material; to Enid H. Bell for critical reading and editorial suggestions; to Janice I. Scott and Ruth Z. Talovich for excellent drafting services; to John S. Shelton, Roland van Huene, Pierre St. Amand, Robert C. Frampton, and Malcolm M. Clark for help in procuring illustrations; to Kathleen G. Nelson for proof reading; and to Joe and Clem Frindt for layman reactions. Colleagues in the Division of Geological and Planetary Sciences of the California Institute of Technology have provided many bits of geological information. Bruce T. Sharp assisted in the construction of field trip guides; and Frank and Irene Goddard, Mack and Enid Bell, and Don and Eileen Burnett provided helpful evaluations of them.

CHAPTER

1

Background

Most of us take the ability to read written material for granted, forgetting how we struggled to learn as small children. Understanding geology through observation of natural features is simply another form of reading. It, too, requires a little knowledge and some practice, but the effort is well worthwhile. Reading of words is essential to safe and comfortable existence in the modern world ("Stop," "Danger," "Women," "Men"), and the reading of nature's geological record makes that existence much more enjoyable.

The principal sources of geological information are rocks and landforms. Both have been around for a long time and have fascinating stories to tell, if we just learn to ask questions in the right way and to listen for the answers carefully. Mostly, it's a matter of being interested and of gradually accumulating experience. If you love natural landscapes, you can soon learn to understand at least some of the history of past events which they have to tell.

The characteristics of landforms are determined principally by three factors: the nature and structure of underlying materials, the surface processes at work, and the stage of development attained in the interaction between processes and materials. A fourth factor important in landscape development, particularly in southern California, is deformation of the earth's surface. In a sense, this deformation is the starting point of most landscape evolution.

MATERIALS

Minerals are the basic units of earth materials. They are solid natural substances of relatively fixed chemical composition and characteristic physical properties such as hardness, luster, and color. The ice in your refrigerator is a mineral, as is the salt you sprinkle on mashed potatoes. Look at table salt with a magnifying glass to see its beautiful little cubic crystals. Many people enjoy collecting minerals, especially gems. Some durable minerals, such as zircon, are prized by geologists because they retain a record of time and events extending over billions of years.

A rock is an aggregate of minerals, either in grains or crystals, although some, such as rock salt or marble, consist of a single mineral. Rocks are the principal constituent of the earth's crust, and they have been used by man from time immemo-

1

rial. One of the first tools our far-distant ancestors used was a rock—to settle an argument with a neighbor, subdue his wife, or kill a wild animal. Rocks are the greatest recorders of history the world has ever known.

You are already familiar with some rocks: the marble of a soda-fountain counter, the granite facing of a bank building, the slate of a blackboard, the sandstone or schist of your fireplace or patio. Rocks come in three principal classes, and you should know their names: *Igneous, sedimentary,* and *metamorphic*. It may help your memory to know that students sometimes speak in jest of "ingenious," "sedentary," and "metaphoric" rocks. Don't expect to recognize all rocks instantly. Even experienced geologists are not above using the cryptic notation *frdk,* for "funny rock don't know."

Igneous rocks were once molten. Upon cooling they crystallized into aggregates of minerals. Sedimentary rocks are secondary, composed of rock or mineral fragments derived from the disintegration and decomposition of preexisting rocks. They can also consist of mineral matter deposited from chemical solutions or of organic substances like plant remains (coal). Metamorphic rocks are also secondary. They are formed by modifications produced in igneous and sedimentary rocks by pressure or heat, or both, sometimes aided by chemical solutions. As a result of such modification new physical characteristics are developed which are wholly unlike those of the original rock. For example, squeezing the living daylights out of shale, a sedimentary rock, compacts it and gives it a special facility for splitting into smooth thin sheets (*rock cleavage*) that makes it the metamorphic rock, *slate*. If the temperature is high enough and significant recrystallization occurs, new minerals are formed and a *schist* is produced. Since igneous and metamorphic rocks are usually composed of aggregates of mineral crystals, they are often spoken of as *crystalline rocks.*

Igneous rocks, other than volcanics, tend to be mostly massive and homogenous. Sedimentary rocks are characteristically layered or bedded. Metamorphic rocks may be either, but in addition, many display an irregular lamination known as *foliation,* most typically seen in *gneiss* (pronounced "nice"). Kent Clark, lyricist laureate of the Caltech campus, once composed a small ditty to gneiss, part of which goes something like this:

> Limestone and coral
> Can never be moral
> They're not gneiss

Here is a simple little rock table with a few common rock names in each class. See the glossary, Appendix B, for more complete descriptions.

IGNEOUS	SEDIMENTARY	METAMORPHIC
Granite	Conglomerate	Slate
Granodiorite	Sandstone	Schist
Diorite	Siltstone	Gneiss
Gabbro	Shale	Marble
Lavas { Basalt Rhyolite	Limestone	Quartzite

GEOLOGICAL STRUCTURES

To most of us the solid earth seems dead because we observe too small a part of it for too short a time. Actually, it is very much alive. It takes energy to keep things living, and the earth gets its energy from disintegration of internal radioactive substances. This produces heat, some of which is dissipated by conduction to the surface and some of which is converted into mechanical energy that bends and fractures rocks. To a geologist, these bends and fractures are *structures;* specifically, they are folds and faults. An up-fold is an *anticline;* a down-fold is a *syncline,* or a "sinkline" in jocular terms (Photo 1-1). Fractures are *joints,* but when significant slippage occurs (or has occurred) along a fracture, it becomes a *fault.* Such movements cause earthquakes, and southern California is richly endowed with both faults and earthquakes. As the late Romeo Martel, dean of earthquake structural engineers, used to say, "California, with all your faults, we love you still, only you don't stay still long enough."

Geological faults of southern California are of two types. Those with principally sidewise or lateral movement, and those upon which the movement has been up or down. We accordingly speak of lateral faults (more properly, lateral-slip faults) or up-down faults.

If you stood on one side of a fault when lateral slip was occurring, the other side would appear to be moving past, either to the right or left. Think about this for a moment and you will realize it doesn't make

Photo 1-1. Folds in mid-Tertiary beds west of parking lot at Calico, field trip Segment C, syncline on left, anticline on right.

any difference which side you stand on, the sense of displacement appears the same, either right or left. Hence come the terms *right-lateral* and *left-lateral* fault. Both sides of a fault may actually have moved in the same direction but one more than the other. In speaking of up-down faults, we need to say that one side has gone up or down relative to the other because both may actually have gone up or down, again, one more than the other. Finally, the fracture surfaces (planes) of some faults are so gently inclined that they are closer to horizontal than vertical. These are called *thrust faults* when one side is shoved over the other.

California's greatest fault is the San Andreas (Photo 2-8). It slashes 650 miles across the state from the Mexican border to Cape Mendocino, 225 miles north-north-west of San Francisco (Figure 1-1). This

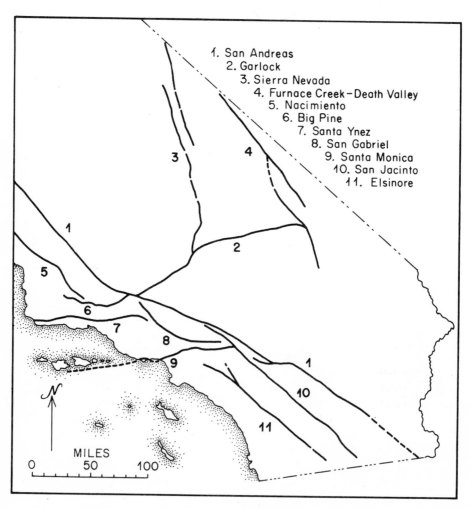

1. San Andreas
2. Garlock
3. Sierra Nevada
4. Furnace Creek – Death Valley
5. Nacimiento
6. Big Pine
7. Santa Ynez
8. San Gabriel
9. Santa Monica
10. San Jacinto
11. Elsinore

MILES
0   50   100

Figure 1-1. Generalized fault map of southern California.

is a classic example of an active, northwest trending, right-lateral fault. Others of this type in southern California are the San Jacinto, Elsinore, and probably the Newport-Inglewood (Figure 2-7). Geological relationships suggest that the total right-lateral displacement on the San Andreas has been tens and possibly hundreds of miles. Displacements on the San Andreas over the last 100 years average out at about 2 inches per year. Since Los Angeles is west of the fault and San Francisco is mostly east of it, these two cities get a little closer together each time the fault slips. The San Andreas or its associated branches are responsible for three of the major interior gateways to the Los Angeles region, namely San Gorgonio, Cajon, and Tejon passes. Many of you have crossed the San Andreas dozens of times.

A different type is represented by east-west faults with predominantly up-down movement, such as the Foothill (Figure 2-7) at the south base of the San Gabriel Mountains. The north side of this fault has moved relatively up creating the steep south face of the San Gabriel Mountains. This face or *scarp* has a vertical relief of 7000 feet in the vicinity of Cucamonga and Etiwanda peaks, something not mentioned by Jack Benny in his many references to Cucamonga. The Banning fault along the south base of the San Bernardino Mountains is a member of this same family.

There are thousands of smaller faults in southern California that do not correspond in trend or sense of displacement with these major types. Furthermore, the Garlock (Photo 3-10), a huge fault extending from Tejon Pass near Lebec to the south end of Death Valley, bears east-northeast and has a large *left* lateral displacement. It belongs in a separate class.

Nearly all sections of sedimentary rocks, more than a few thousand feet thick, in southern California are tilted or folded. They are also faulted, as faults have no respect for rock types. Only rarely in this domain of ours do you see near-horizontal bedding, and when seen, it is almost invariably a mark of geologically young deposits.

Many of our topographic prominences such as Signal Hill, Domingues Hills, Santa Monica Mountains, and Wheeler Ridge (Photo 2-9), among others, are primarily the result of folding. We recognize up-folds (anticlines) more readily than down-folds (synclines) because the former make ridges or ranges, and the latter make valleys that become filled with rock, sand, and dirt (*alluvium*). San Fernando Valley is basically a large synclinal structure lying north of the anticlinal Santa Monica Mountains. As a structure it extends farther down into the earth than the Santa Monica Mountains extend up into the sky. It is easy to think too much of our mountains and to forget that our valleys are also geologically significant (Figure 2-3).

Massive igneous and metamorphic rocks are subject to the same forces that cause sedimentary beds to fold. Usually they respond by faulting, but in some instances they warp. This means that they bend or tilt gently on a broad scale. Warping is an important geological phenomenon, but the results are often hard to recognize. If it weren't for the layering in sedimentary rocks, it would not be obvious that they had been folded.

Southern California is different from much of the rest of the nation in that our folding and faulting have occurred so recently that much of the topographic relief is due directly to such movements. High standing areas have been folded, warped, or faulted upward, and low standing areas the reverse. This is not true in many parts of the country where the deformation occurred long ago. There the present relief is due primarily to differential erosion of hard and soft masses of rock.

It is the recency and variety of geological structures that endows California with its unusual terrain, including the highest and lowest spots of our contiguous states. California rocks are not all that different from those in some other parts of the country, and processes acting on these rocks are about the same as elsewhere. We simply live in a geologically dynamic area, and it shows. In subtle, psychological ways our youthful geology may be partly responsible for the high basal metabolism and mobility of our west-coast culture.

PROCESSES

So much for materials and structures. Now what are the processes working on them to carve out the landscape? There are a number, but they do not all act in concert or with the same degree of effectiveness in different environments. Basically, what we are dealing with is the interaction of the solid earth with the atmosphere, hydrosphere, and biosphere, including man. Man modifies the earth's surface with all of his chicken scratches. He just hasn't been here long enough to carve out a Grand Canyon. However, for his size he is an efficient dirt mover.

Weathering. If you leave your car sitting out in the open long enough, it will start to rust and disintegrate. That's *weathering,* the reaction of solid substances with the atmosphere and biosphere. Well, rock surfaces have been exposed to the elements for hundreds of thousands of years. Conse-

quently, essentially all of them display some degree of weathering. Many geologists didn't know what truly fresh unaltered rock looked like until they saw chunks of moon rock. Weathering is a pervasive, night and day, omnipresent process. It works slowly, but over long periods of time it is very effective. Without it you and I wouldn't be here, for weathering produces one of our necessary natural resources, soil. Given time and proper conditions, weathering will make a fine soil out of a solid mass of granite. The chemical and physical changes produced in rocks by weathering play a major role in landscape development by making rocks susceptible to erosion.

Mass Movements. Once rocks have been softened and disintegrated by weathering, the debris starts to move. Initial downslope movements are caused by gravity unaided by other agents of transport. The material simply creeps, slides, rolls, or flows downslope. This is mass movement.

Some types of mass movement are so slow as to be imperceptible to short-time observation. If you planted a statue of Venus de Milo on a soil-mantled hillside, you would probably find over a period of 10 years that the good lady would tilt a bit and move downslope a foot or two. This behavior results from the insidious process of *creep* which prevails to some degree on practically every debris-mantled slope.

Earth flows (Photo 1-2) and *mudflows* are forms of mass movement which occur episodically and usually at an easily observable rate. *Landslides* move with even greater velocity and are fully capable of engulfing anyone in their path. Some, like the rock-fall slide (Photo 1-3) that took place near Blackhawk Canyon on the north face of the San Bernardino Mountains many thousands of years ago, probably attained a velocity in excess of 100 miles an hour.

Photo 1-2. Earthflow in central Puente Hills, viewed northeastward, San Gabriel Mountains in background, Old Baldy upper left. (Photo by John S. Shelton, 720).

Photo 1-3. Breccia lobe formed by Blackhawk Canyon landslide, north side San Bernardino Mountains. Toe of lobe is 4.5 miles from base of mountains. (Photo by John S. Shelton, 2459).

This lobate tongue of broken rock traveled 5 miles out over the desert floor at the east end of Lucerne Valley before coming to rest.

Running water. The most effective agent of erosion is running water. Any high-standing land mass is subjected to its attack, except in unusual places, such as Antarctica, where the water is frozen. Running streams carve valleys, canyons, gorges, arroyos, barrancas, and gullies, all integral parts of our landscape. It's easy to underestimate the effectiveness of a stream until it goes into flood. Then skeptics quickly become believers, for streams are terrifyingly effective at times of flood. Even on southern California deserts, water is the principal agent of erosion and transport because it is not impeded much by vegetation; hence flash floods are common.

We think of running water too much as a destructive agent. Through deposition it also acts constructively. Some of the most heavily populated parts of the southern California landscape, alluvial fans (Photo 3-8) and alluvial plains, have been constructed by stream deposition.

Wind. Wind is localized in its geological effectiveness, but where it works, it's great. Wind erodes, transports, and deposits. Wind-blown sand is a highly effective agent of abrasion, as anyone emerging from a Coachella Valley sandstorm with a frosted windshield can testify. In addition to blasting windshields, wind-blown sand undercuts powerline poles and etches and polishes stones resting on the ground, converting them to *ventifacts* (literally— 'wind-made'). They are not wind-made, of course, only wind-shaped.

Most agents transport material downhill. Wind is one of the few agents which can and does carry material uphill. In deserts, winds carry sand for tens of miles before piling it up into huge dunes hundreds of feet high (Photo 3-4). The shapes and forms associated with sand dunes are some of the loveliest in nature, especially when seen in low evening or morning lighting. That's the witching time in dunes.

Glaciers. The San Bernardino Mountains harbored seven small valley glaciers clustered in the San Gorgonio Peak area during the last ice age, about 15,000 years ago. However, the Sierra Nevada displays the erosional and depositional products of mountain glaciers at their best. The broad U-shaped valleys, glacial steps, waterfalls, hanging valleys, polished rock surfaces, bedrock-basin lakes, and huge piles of rocky debris known as *moraines* left by the glacier are as well displayed there as anywhere in the nation.

Shoreline processes. The principal source of energy at the earth's surface is radiation from the sun. About 70 per cent of the earth is covered by oceans which intercept the major share of this solar radiation. Oceans are therefore great pools of energy, some of which they expend along their contact with land, the seashore.

Owing to the constant attack of waves and currents, the seashore is geologically dynamic. There's a lot of erosion, transportation, and deposition going on, and the scene is ever changing. Sea cliffs retreat at rates measured in many feet per year, and beaches change aspect with every storm, and regularly with the season. Winter storms move sand off California beaches leaving piles of boulders. In summer, waves and currents bring sand back burying the boulders and restoring the beaches. The balance of sand supply and sand movement along beaches is a delicate thing, as man has learned time and time again to his sorrow when he has interfered with shoreline processes. Our current damming of flood waters in streams on land is reducing the amount of sand supplied to the ocean. In time this program is going to adversely affect our beaches.

GEOLOGICAL TIME SCALE

Don't be overawed by the carefree way geologists speak of millions and billions of years. We humans are on this good earth for such a few ticks of the geological clock, which started running 4.5 billion years ago, that it's hard for us to put geological events into proper perspective. If the entire span of geological time were compressed into one year, and we imagined the present to be exactly midnight of December 31, one of my associates, George Clark, calculates that the Declaration of Independence was signed 1.4 seconds ago, Columbus discovered America 3.3 seconds ago, Christ was born about 14 seconds ago, and a hundred seconds ago ice covered much of Europe and North America and our Stone Age ancestors huddled in caves. Man appeared on Earth at about suppertime, 6 P.M. this evening, and living creatures first crawled out of the water back in the last week of November. This should help you appreciate the inadequacy of our own personal experiences with the earth, its activities and processes because of the short time involved. That's why geologists are so interested in talking with rocks 2 and 3 billion years old. Those rocks are wise in the ways of the world.

For our purposes relative age relationships are also important. Geologists have built up a relative time chart filled with curious names; a simplified version is supplied in Appendix A. Don't try to memorize this chart, but look at it from time to time as you come across geological-time names in these pages. Gradually you will learn that something classed as Cenozoic is relatively young and that Precambrian rocks or events are relatively ancient. The geological time scale is your chart for navigating the vast seas of geological history. Don't hesitate to use it.

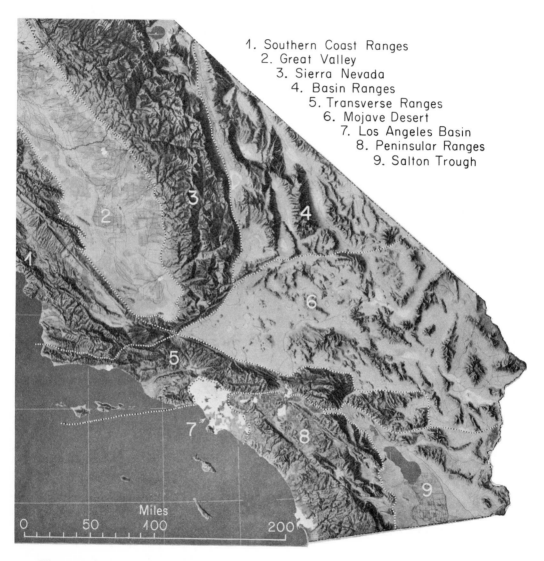

1. Southern Coast Ranges
2. Great Valley
3. Sierra Nevada
4. Basin Ranges
5. Transverse Ranges
6. Mojave Desert
7. Los Angeles Basin
8. Peninsular Ranges
9. Salton Trough

Figure 2-1. Natural provinces of southern California. Base Map © 1961 Jeppesen and Co., Denver, Colorado, USA. All rights reserved.

10

⬭ 2 ⬭

Natural Provinces
of Southern California

Man has mental limitations, and nature is infinitely complex. To deal with this situation man invents classifications. Nature does not classify trees, flowers, and rocks— we do, so that we can deal with them in a reasonable fashion. In order to facilitate geological understanding, southern California is divided into nine provinces (Figure 2-1) even though the procedure does some violence to natural relationships. A natural province is a region with characteristics which distinguish it from other regions. It can be defined on the basis of vegetation, climate, topography, or the color of jack rabbits. We use geology.

The U. S. Geological Survey sells a dandy small-scale geological map of California for only 25 cents. It shows natural provinces, major faults, and a generalized distribution of rock types. *See* Appendix C on directions for obtaining a copy.

TRANSVERSE RANGES

New arrivals, and even a few "aborigines" (residents of more than 10 years), in the greater Los Angeles-Santa Barbara areas,

easily become confused on compass directions, and with good reason. "Up the coast" is commonly taken to mean north, but it's not north in this province. In traveling a straight course from San Bernardino to Santa Barbara, one ends up 137 miles west and only 27 miles north. To go north from Los Angeles you travel almost directly into the interior toward Mojave, crossing east-west mountains in the process.

CHARACTER AND DIMENSIONS

This east-west landscape defines a geological province, the Transverse Ranges, so named because they are crosswise to the more prevailing northwesterly fabric of California. This characteristic is established by faults and folds which control the trend and shape of mountains, valleys, and the coastline. The cause of this unusual orientation has long puzzled geologists, but recent explorations on the Pacific Ocean floor reveal a number of huge structural lines extending directly eastward toward the coast of North America. One of them, the Murray fracture zone, reaches the continent in the region of the Transverse Ranges. These ranges may thus represent the landward influence of a sea-floor structure.

Topographically this province embraces a rugged chunk of country featuring high, rough mountain masses and long, narrow, intervening valleys. Many of the higher peaks of southern California, outside the southern Sierra Nevada, are in this province — San Gorgonio which is 11,502 feet and San Antonio (Old Baldy), 10,064 feet, are two.

The principal ranges and valleys of this province are identified in Figure 2-2. The Channel Islands are included since they are but a partly submerged westward extension

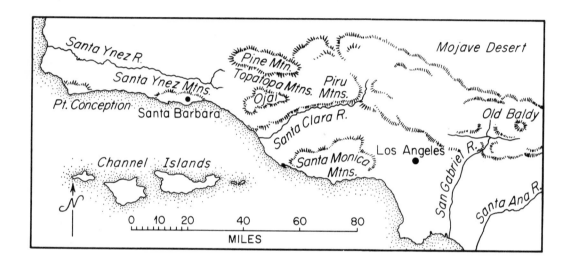

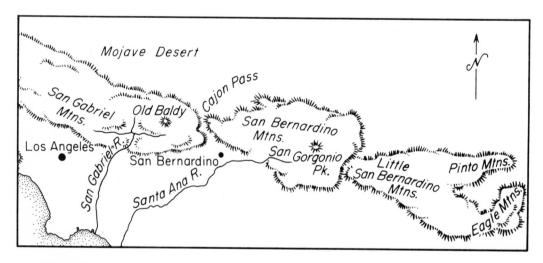

Figure 2-2. Map of Transverse Ranges province, west half above, east half, with overlap, below.

of the Santa Monica Mountains. The Transverse Ranges province is nearly 300 miles long, extending from Point Arguello, 55 miles west of Santa Barbara, eastward to Eagle Mountains in the desert.

ROCKS

It is convenient to think of this province as divided into eastern and western parts by the Golden State Freeway between Los Angeles and Bakersfield. In the west, sedimentary rocks predominate; in the east, older igneous and metamorphic rocks are the rule. Volcanic rocks are locally represented in both parts. The Santa Ynez, Topatopa, Piru, and Pine Mountain ranges (Figure 2-2) display superb successions of sedimentary rocks, aggregating tens of thousands of feet in thickness. The San Gabriel, San Bernardino, and other ranges farther east are a mixture of igneous and meta-

morphic rocks, including some of the most ancient in southern California, roughly 1.7 billion years old. The intervening valleys are underlaid by thick accumulations of sedimentary beds or by crystalline rocks buried under unconsolidated deposits of sand, gravel, and dirt, called *alluvium.*

STRUCTURE

An east-west alignment of folds and faults produces mountains and valleys of that orientation. These topographic features in turn control lines of communication (highways) and the location and shape of our populated areas. Folds are most prominent in the west where the sedimentary section is thickest. The principal ranges there are up-folded, that is, *anticlinal,* and the valleys are down-folded, or *synclinal,* as shown by the cross section of Santa Clara Valley (Figure 2-3). This western area is

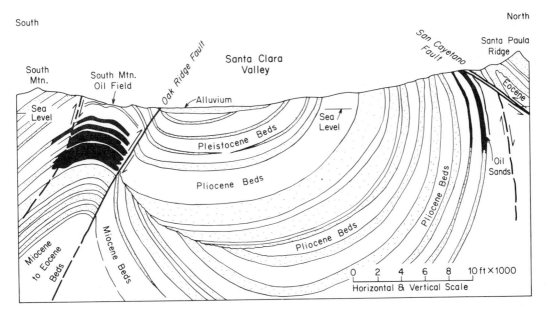

Figure 2-3. Geological cross section of the Santa Clara Valley.

not without its faults, but folds dominate the scene.

In the east, faults are the principal structures. Many, like the Foothill fault along the south side of the San Gabriel Mountains, trend east-west and are steeply inclined to the north with the north side relatively uplifted. The San Gabriel is a different type of fault which cuts through the heart of those mountains, determining the course of the east and west forks of San Gabriel River before flaring off northwestward at Castaic. The San Gabriel fault shows evidence of many miles of right-lateral slip.

The principal dissonant note in the east-west fabric of the Transverse Ranges is the San Andreas fault, and its close relative, the San Jacinto, which slash across the province in a northwesterly direction through the Cajon Pass-Lytle Creek region. These are large active faults with many, many miles of right-lateral displacement. The San Gabriel and San Bernardino mountains would make a continuous range if they were not sliced apart by these gigantic ruptures. Flying out of Los Angeles east-bound via Cajon Pass, you can often see off the left wing of your plane the great gash of the San Andreas cutting discordantly across the terrain.

RESOURCES

Much oil has been produced from the western part of the Transverse Ranges where the many folds, and some associated faults, provide traps for its accumulation. Petroleum geologists long ago discovered that oil, which is lighter than water, tends to migrate upward into anticlines where it accumulates beneath the curving cap of some inpervious bed. If the impervious cap is cracked by faults and fractures, oil may seep through to the surface. This has been happening for thousands of years in the Santa Barbara Channel, and such leaks were basically the reason for the oil slicks associated with drilling operations in the area in 1969.

If oil is king in the west, iron is queen of the east, where the Eagle Mountains supply much of the ore for Kaiser's Fontana mills. In between, limestone is quarried for cement, some building stones are produced, sand and gravel are exploited, and a little placer gold has been recovered from the East Fork of San Gabriel River. A water-laid accumulation of rock debris, usually sand and gravel containing fragments of precious mineral, is a placer deposit.

SPECIAL FEATURES

Cabrillo Peninsula. A few million years ago the structural uplift, represented by Santa Monica Mountains, extended far westward as a long narrow peninsula converting the present Santa Barbara Channel into a bay (Figure 2-4). This peninsula, named in honor of the great Portuguese explorer, Juan Rodriguez Cabrillo, encompassed the present Channel Islands of Anacapa, Santa Cruz, Santa Rosa, and San Miguel, where Cabrillo is supposedly buried. Structurally, these islands are anticlinal with numerous associated faults, one of which on the south side appears to be the far-western extension of the Santa Monica fault (Figure 1-1).

Cabrillo Peninsula was inhabited by mainland plants and animals, including elephants which were then reasonably abundant in California. Eventually, perhaps 1 or 2 million years ago, the Cabrillo Peninsula subsided leaving only its higher points projecting as the present islands. The poor elephants were isolated, and in the course of time evolved into a race of pygmies standing

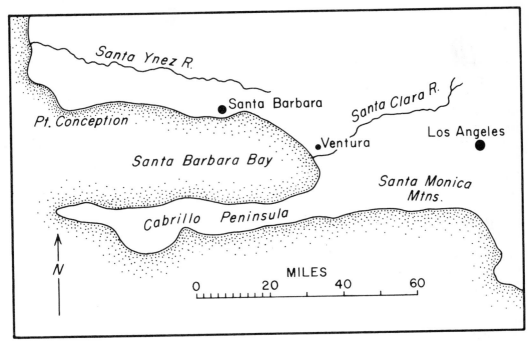

Figure 2-4. Map of Ancient Cabrillo Peninsula.

only 6-8 feet high, compared to a normal elephant height of 10-13 feet. This change is attributed to insular isolation, but just how or why isolation on an island breeds smaller animals is not wholly clear. Maybe they had to huddle in caves, and only the smaller ones could get in.

Sespe Formation. To geologists, a *formation* is a body of rock of consistent characteristics formed within a discrete span of time under specific conditions. One distinguishing feature of the Sespe Formation is its red color. It is not all red, but those parts that are, like red-headed girls, attract attention. The Sespe is also a land-laid formation, that is, it was deposited as a series of gravel, sand, and mud layers on land rather than in the sea. This distinguishes it from thick sequences of sedimentary beds above, below, and alongside that are of marine origin. The Sespe represents nearly 20 m.y. of time, embracing the interval from late Eocene to early Miocene (*see* Appendix A).

The Sespe is not restricted to the Transverse Ranges, but that is its principal abode. Remnants of it are preserved in many places between Point Conception and the Santa Ana Mountains. They are seen in lower Topanga Canyon, west along the south face of the Santa Monica Mountains, on the Channel Islands, in Simi Valley, the Las Posas, in South Mountain opposite Santa Paula, and on the upper and lower parts of Sespe Creek, whence comes the name. When traveling up or down the Santa Clara Valley, look north to the mouth of the Sespe Canyon as you cross the Sespe Creek bridge just south of Fillmore. If the light is right, you will see dark-red sandstone beds east of the canyon mouth. Those are Sespe beds, and large reddish boulders in the stream bed are samples of the rock.

The Sespe Formation was deposited in a large land-locked basin bounded by relatively high mountains. The valleys of this basin must have been well watered because they were inhabited by large rhinoceros-like beasts and other heavy vegetational browsers. Picture a lushly vegetated basin extending 150 miles east-west across southern California and imagine what a scene it made when populated with lemur monkeys, primitive dogs, hyena-like carnivores, camels, horses, primitive pig-like cud chewers (oreodonts), deer-like animals, tapirs which looked like an early form of small stocky horse with a pendant snout, and a variety of large rhino-like animals tromping around the premises. There were even a few primitive saber-toothed cats and lots of small rodents and rabbits. Titanotheres (titanic mammals), sort of a cross between a horse and a rhinoceros, also stalked the scene. Such was this land of ours in Sespe time; it looked more like a part of present-day Africa.

PENINSULAR RANGES

The Peninsular Ranges in the southwest corner of southern California constitute a land of gems, telescopes, and *batholiths*. Don't recoil from that word—*bathos* = 'deep', *lithos* = 'rock'; it's that simple. A batholith is an igneous intrusive body of large size which cooled so slowly that the molten material had time to crystallize into a coarse-grained rock. This occurs most easily at depth. The southern California batholith finds its home in this province, the world's largest optical telescope is on Palomar Mountain in the Agua Tibia Range, and the Pala area was once one of the great gem collecting areas of the nation. The

Peninsular Ranges actually belong more to Mexico than the U. S. since they extend the entire 800 miles of peninsular Baja California, whence comes the name. We claim only a small fraction of the province up here in gringo-land, but that's the part under consideration.

CHARACTER AND DIMENSIONS

This province has a distinct, but not overpowering, northwest grain expressed by its higher mountains and longer valleys (Figure 2-5). It also includes a lot of hilly country without strong linear pattern, such as the area around Fallbrook. There's a hill here, a vale there, and the roads and ranches are comfortably dispersed about the landscape. This local lack of linear geometrical pattern reflects the homogeneity of the underlying batholithic rocks.

The Peninsular Ranges merge northward into the Los Angeles Basin, and the northwest grain eventually terminates against the east-west Transverse Ranges. The Penisular Ranges are bounded on the east by the Salton Trough. Westward, the province does not end at the Pacific shore, as one might assume, but continues far out under the ocean as a broad submerged continental borderland. The islands of Catalina, San Clemente, Santa Barbara, and San Nicholas are simply the high parts of largely submerged peninsular-type ranges.

The U. S. part of the province extends 130 miles north from the Mexican border. The maximum landbound width is about 65 miles, but including the submerged continental borderland, the width is 225 miles. Some of the larger ranges and major valleys within the landbound sector are identified on Figure 2-5. Hot springs keep things warm here and there, Warner's being the best known.

In gross aspect, the province is a large

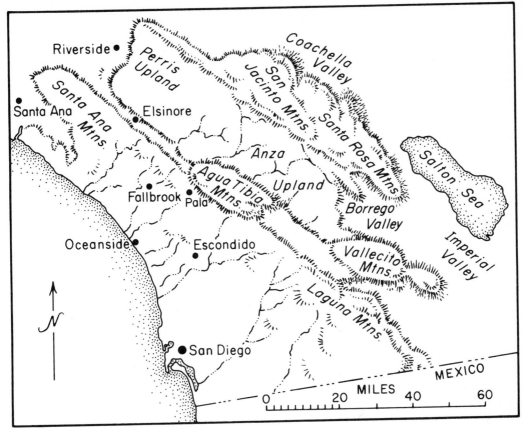

Figure 2-5. Map of Peninsular Ranges province.

block uplifted abruptly along the eastern edge and tilted westward. The highest point, San Jacinto Peak (10,831 ft.), towers more than 10,000 feet above Palm Springs and Coachella Valley. This scarp is as high as the east face of the Sierra Nevada and surpasses in vertical relief the east face of the Grand Tetons west of Jackson Hole, Wyoming.

The mountains and ranges making up the west slope rise steeply and abruptly above adjoining valleys, but much of the upland country has a gentle relief. The area around Palomar Observatory, the west side of the Santa Rosa Mountains along the Palms-to-Pines Highway, and the west flank of the Laguna Mountains farther south display large areas of gentle upland. The flat terrain around Perris and March Field is similar but occurs at a lower level. Faulting and homogeneous igneous rock play major roles along with erosion in producing this steep-sided, gentle-topped topography.

Much of the coastal margin has one or more wide flat benches upon which the highways and coastal settlements are situated. These are *marine terraces;* the flat treads represent uplifted sea-floor platforms, and the steep slopes between are old sea cliffs.

ROCKS

If you dote on relatively coarse-grained, homogeneous igneous rocks, and if you like the type of soil and landscape they yield,

17

this is your province. Rocks of the southern California batholith are dominant. Although formed at depth, they have subsequently been exposed at the surface by erosion. However, since these rocks had to be intruded into something older, it is not surprising to find sheaths of metamorphic rock around the margins of intrusive bodies, and within them metamorphic remnants. The oldest of these metamorphics are schists, quartzites, and coarsely crystalline marbles. A somewhat younger group of prebatholithic rocks appears originally to have been shales and volcanics.

The batholithic rocks consist of separate intrusive units ranging in composition from gabbro (dark) to granite (light). They are accompanied by a host of *dikes,* narrow sheet-like igneous bodies intruded into cracks. One dike rock, called *pegmatite,* is particularly important. A pegmatite is often of granitic composition, and it is very coarse-grained with individual crystals measured in inches, sometimes feet. Pegmatites seem to be formed by the last vapors and fluids given off by a cooling and crystallizing igneous body. Rocks of the southern California batholith are 90-100 m.y. old.

Younger rocks found in the province are largely sedimentary, partly marine and partly terrestrial, and range in age from late Cretaceous (80 m.y.) to Pleistocene (2 m.y. or less). Marine rocks are exposed mostly in the Santa Ana Mountains and in the belt of hills along the coast from Corona Del Mar to San Diego. Terrestrial deposits were laid down in inland basins; the Borrego Badlands expose this type of accumulation. Volcanics of about mid-Miocene age (20 m.y.) appear in some places, including Catalina and San Clemente islands, the latter being largely volcanic. Still younger lavas are preserved south of Lake Elsinore.

STRUCTURES

The northwest grain of the Peninsular Ranges is caused primarily by faulting. Folding is locally significant in the Santa Ana Mountains and coastal hills, and parts of the province have probably been broadly warped. Major faults are the San Jacinto and Elsinore (Figure 1-1), and both are active. These are complex structural zones accompanied by a number of parallel, individually named satellitic faults, particularly southeastward. A similar pattern of faults controls the topography of the offshore continental borderland.

The San Jacinto and Elsinore faults, like the San Andreas, have experienced much right-lateral displacement. Many of the other faults show considerable vertical movement, although the present topographic relief along them is attributed as much to differential erosion of rocks of different hardness on opposite sides of the fault as to direct displacement. In terms of energy released within historical times in southern California, the San Jacinto has been more active than the San Andreas.

RESOURCES

Attempts have been made to mine gold, lead, zinc, copper, nickel, and tungsten in this province, but only the production of gold, largely from quartz veins in the Julian area, has been significant. Some gravels in the region contain small accumulations of placer gold.

The province is more noted for its nonmetallic products. Cement comes from marble bodies in the Jurupa Mountains near Colton and Riverside. Gypsum is mined in the Fish Creek Mountains of Imperial Valley, and building and decorative stones are quarried in the southern California batholith. Clay and some low-grade coal were formerly produced near Elsinore. Roofing granules come from several places, and sand and gravel from many.

SPECIAL FEATURES

Since 1872 the Pala district, near the center of this province, has been one of the more noted gem-mineral collecting areas of the nation. Something approaching a million dollars of gemstones have been commercially produced, mostly between 1900 and 1922, and large quantities of rare minerals and gems have been obained for private collections. The region is particularly noted for its lithium bearing minerals, one of which, spodumene, a lithium aluminum silicate, forms gems with exquisitely delicate pink and lavender tints. Tourmaline, the other common gemstone, is a chemically complex compound which is attractive because of its hardness, high luster, and brilliant green and pink coloration.

These minerals occur in pegmatite dikes of which over 400 have been mapped in the Pala district. They are irregular bodies, commonly lenticular, but occasionally with bulges 100 feet across. Pegmatites tend to be extremely coarse grained, and they contain a wide variety of unusual and highly volatile chemical elements, which is the reason for their exotic mineral composition. Mineral collectors still find interesting "pickings" in the Pala area.

LOS ANGELES BASIN

Los Angeles Basin means different things to different people. To officials of the Air Pollution Control District, it is that region all too often submerged in smog. To professional geologists, it signifies a basin on the sea floor in which a great thickness of mud and sand accumulated in times past. For purposes of this booklet, it is that part of our land extending south from the foot of San Gabriel Mountains to the sea and southeast from Santa Monica Mountains to Santa Ana Mountains and San Joaquin Hills. (Figure 2-6).

Many geologists include Los Angeles Basin within the Peninsular Ranges because it has the same northwest structural grain. We handle it separately and accord it a more detailed consideration because it is the abode of so many people and because none of the geological trip guides of this booklet deal with the Los Angeles Basin.

CHARACTER AND DIMENSIONS

The Los Angeles Basin is divided into a northern one-third and a southern two-thirds by the Puente-Repetto hills (Figure 2-6). South of these hills is a lowland sloping gently toward the sea—a coastal plain. North is San Gabriel Valley, an alluvium filled basin encircled by hills and mountains.

The coastal plain rises from the sea to elevations of a few hundred feet with a few scattered hills and mesas projecting 100-200 feet above its level; Baldwin, Dominguez, and Signal hills are examples. Palos Verdes Hills at the southwest edge interrupt the otherwise smooth transition of the coastal plain to the sea. They are like an offshore island captured by seaward growth of the land. The Los Angeles Basin is enclosed on its landward sides by mountains rising abruptly thousands of feet. The basin surface irregularly penetrates these bounding hills and mountains like a sea encroaching upon an irregular mountainous coastline.

Three of southern California's larger streams, the Los Angeles, San Gabriel, and

Santa Ana rivers traverse the basin. They are in large part responsible for the alluvium that now mantles its surface, and they also cut right across hills that lie athwart their path.

The long axis of the basin extends northwesterly fully 50 miles from the San Joaquin Hills near Laguna to the Santa Monica Mountains. The width from Palos Verdes Hills to San Gabriel Mountains approaches 35 miles.

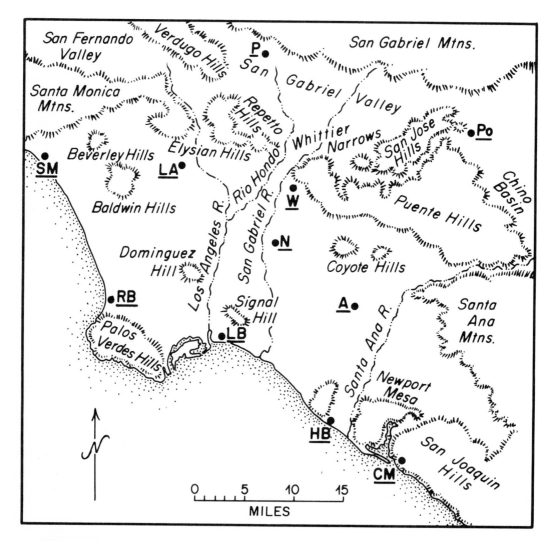

Figure 2-6. Map of Los Angeles Basin; A—Anaheim, CM—Corona Del Mar, HB—Huntington Beach, LA—Los Angeles, LB—Long Beach, N—Norwalk, P—Pasadena, Po—Pomona, RB—Redondo Beach, SM—Santa Monica, W—Whittier. Modified from R. F. Yerkes *et al.*, 1965, USGS Prof. Paper 420-A.

ROCKS

The surface of Los Angeles Basin is largely covered by stream-laid sand, gravel, and silt which effectively hide the underlying bedrock. However, the basin is as full of holes as high-grade Swiss cheese, these holes being the tens of thousands of wells drilled in search of oil and water. From them, and from much subsurface geophysical exploration by oil companies, a lot is known about the thickness and nature of rocks beneath the surface. However, to see these rocks in outcrops one must repair to hilly areas within the basin or to mountainous areas bordering it. The rocks of Los Angeles Basin comprise three principal groups: (a) old crystalline basement rocks, (b) pre-basin sedimentary rocks, and (c) materials filling the basin, called basin sediments or basin fill.

The basement rocks are of two types, called for convenience, the eastern and western complexes. The eastern complex consists of igneous and metamorphic rocks of the types seen in the Peninsular and Transverse ranges. The western complex consists primarily of an unusual metamorphic rock, the Catalina schist. It is exposed on Catalina Island and in a small area high on the northeast flank of the Palos Verdes Hills. The Catalina schist is unusual in that some of its minerals are abnormally high in sodium which gives them a peculiar bluish color. Exposed in sea cliffs around Laguna is a coarse bouldery deposit containing large angular fragments of this schist. The location, age, and structure of this deposit suggests that roughly 20 million years ago a high landmass lay just off the present coastline shedding coarse rocky debris into the Laguna area.

Younger than the basement rocks, but still older than the basin fill, are sedimentary deposits ranging from late Cretaceous (80 m.y.) to lower Miocene (25 m.y.). They are layers of conglomerate, sandstone, and shale aggregating a total thickness of nearly 17,000 feet. Included is our old friend, the Sespe Formation. The Los Angeles Basin did not exist as a separate feature when these deposits were laid down.

Subsidence of the Los Angeles depositional basin started about 20 million years ago in mid-Miocene time, and was accompanied by the extrusion of volcanic material. The Topanga Formation, the deposit first laid down in the basin, locally contains as much as 3000 feet of volcanic rocks and associated near-surface intrusives within its total thickness of 10,000 feet. It's fun to reflect that not so long ago our backyard was dotted with active volcanos spouting smoke, fumes, ash, fragmented ejecta, and red-hot lava over the surroundings. These volcanic rocks can be seen today around the edge of the basin, in Palos Verdes Hills, Griffith Park, or Puddingstone State Park of the Covina Hills.

A little uplift and erosion occurred before renewed subsidence led to additional deposition of thick sections of Miocene, Pliocene, and Pleistocene sedimentary beds, all of marine origin. In places, the Pliocene sandstone, siltstone, shale, and conglomerate alone have an aggregate thickness of 14,000 feet. Some of these deposits are exposed in the Repetto Hills (Figure 2-6) which give their name to the Pliocene Repetto Formation. The large brickyard along the Long Beach Freeway where the freeway cuts through these hills makes bricks from the clay-rich shales of the Repetto Formation.

The earliest Pleistocene deposits are marine, and some contain abundant fossil sea shells, particularly in Palos Verdes Hills.

However, as the basin filled, the surface rose above the sea and terrestrial deposition followed the receding shoreline southward. One of the most unusual of these Pleistocene terrestrial deposits is the tar-pit accumulation of Rancho La Brea. Here bones, representing the amazing animal life of southern California 12,000 to more than 40,000 years ago, are preserved in tar. By all means, visit the exhibits of Rancho La Brea fossils in the Los Angeles County Museum in Exposition Park and the restorations at the fossil pits in Hancock Park. It is pointless to write about the saber-toothed tigers, giant wolves, ground sloths, elephants, buffalo, lions, and the birds of Rancho La Brea when you can see their restored remains for yourself.

STRUCTURE

Los Angeles Basin is a huge downfold (syncline) with a central line (*axis*) extending northwest from Santa Ana to Beverly Hills (Figure 2-7). The next time some world traveler regales you with stories of towering Himalayan peaks just say, "Let me tell you about the basement relief of the Los Angeles Basin." The lowest part of the basement surface lies between South Gate and Downey at 31,000 feet below sea level. The top of Mt. Wilson, only 20 miles north and composed of similar basement rock, approaches 6000 feet above sea level. Total basement relief is thus 37,000 feet. If all the sedimentary fill were cleaned out of the basin, the topographic relief would exceed that of the Himalayas at their highest point, Mt. Everest, by nearly 7000 feet. It would be a varied topography to boot. As you can see from Figure 2-7, the Los Angeles Basin is replete with folds and faults. Each of these structures is represented by a major feature of relief, thousands

of feet high, on the underlying basement floor.

These structures bear northwesterly, which is why many geologists include Los Angeles Basin within the Peninsular Ranges. Indeed, some of the structures actually continue from the Peninsular Ranges into the basin, but note how they are all cut off sharply on the northwest, at the junction with the Transverse Ranges, by the Santa Monica-Raymond fault. Faults determine the boundaries of Los Angeles Basin on two other sides, namely, the Palos Verdes Hills fault on the southwest and the Foothill on the north. Faults also define units or blocks within the basin, the central block, for example, being bounded on the northeast by the Whittier fault and on the southwest by the Newport-Inglewood fault.

The Newport-Inglewood fault zone is a complex structure with a northwesterly trace marked by a succession of mesas and hills extending from Newport Bay to Beverly Hills, of which Signal, Dominguez, and Baldwin hills are the more prominent. It produces a vertical displacement in the underlying basement of nearly 4000 feet and separates the western and eastern basement complexes, suggesting that this may be an old feature of truly major displacement. In the overlying sedimentary rocks the fault zone consists of a sequence of short parallel overlapping linear segments, and the displacement decreases upward so that the youngest beds are offset only 200-300 feet. This indicates that movement on the fault has occurred repeatedly and in small increments. Old rocks record the cumulative sum of these displacements, but young rocks show only the latest movements. Other structures in the Los Angeles Basin, folds as well as faults, suggest a similar history, that is, initiation at about the time of original basin subsidence 20 million years

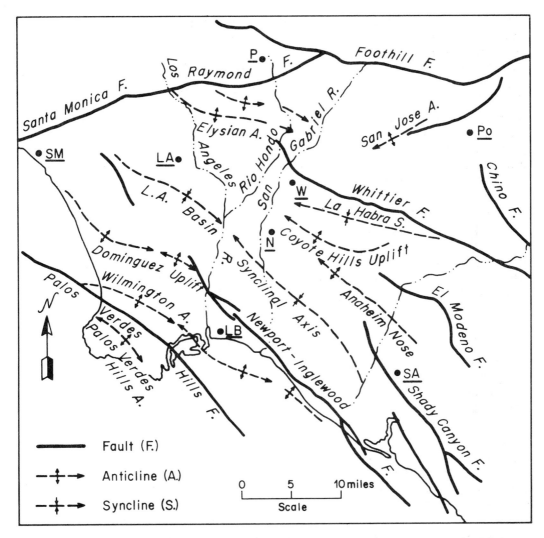

Figure 2-7. Map of geological structures in the Los Angeles Basin; LA—Los Angeles, LB—Long Beach, N—Norwalk, P—Pasadena, Po—Pomona, SA—Santa Ana, SM—Santa Monica, W—Whittier. Modified from R. F. Yerkes *et al.*, 1965, USGS Prof. Paper 420-A.

ago and subsequent intermittent activity up to the present. Although possibly old, the Newport-Inglewood fault zone is anything but dead, as attested by the Capistrano earthquake of 1812 and the disastrous Long Beach shock of 1933. Geological relationships suggest considerable right-lateral dis-

placement along the Newport-Inglewood fault, so it may be a member of the San Andreas family.

Some structures extend from the Peninsular Ranges into the Los Angeles Basin, Shady Canyon and El Modeno faults at the southeast corner (Figure 2-7) for example.

Whittier fault also appears to be essentially a continuation of the Elsinore fault, a major structure of the Peninsular Ranges. The Whittier fault extends along the south base of the Puente Hills and dies out west of Whittier narrows in the Repetto Hills. Like the Elsinore, it exhibits significant right-lateral movement, but it also displays much vertical displacement. The actual movement has probably been oblique, that is, partly up and partly lateral, and it has not been inconsequential, totaling at least 15,000 feet.

San Gabriel Valley lies between Puente-Repetto Hills and San Gabriel Mountains. The latter have been formed by large and recent uplift along the east-west Foothill fault. Splitting off from the Foothill fault in the Arcadia-Monrovia area is the Raymond fault. The Huntington Hotel and the Huntington Library are perched atop a young scrap marking the trace of this fault through the Pasadena-San Marino area. Horse racing fans may enjoy knowing that the hillside stretch of the turf course at Santa Anita comes obliquely down the face of the Raymond scarp. The fault runs westward to join the Santa Monica fault which extends along the south side of the Santa Monica Mountains, past Malibu and Corral Canyon, and far westward along the south side of the Channel Islands (Figure 1-1).

Hilly areas within the Los Angeles Basin are mostly anticlines, trending principally northwesterly and plunging in that direction. Take this booklet, fold it into an anticline, and then tilt it so the crest is no longer horizontal. Now you have a plunging anticline. The arrow at the end of a line representing a fold on Figure 2-7 indicates the direction of plunge. Some folds are doubly plunging, that is, they plunge northwest in one place and southeast in another,

Palos Verdes Hills anticline being an example. The Los Angeles Basin synclinal axis also plunges from both directions toward its lowest spot under the junction of Rio Hondo with the Los Angeles River near South Gate (Figure 2-7).

A geologically interesting aspect of most folds in the Los Angeles Basin is their youth. They are so young that the uplifted areas have the configuration of the structure itself. Although dissected by gullies and canyons, a doubly plunging anticline still looks in profile like a doubly plunging anticline. This is not true of structures in many other parts of the world. Earthquakes and the measurement of geodetic markers suggest that deformation is still going on in Los Angeles Basin.

Some of you may be aware that the area around Anaheim has sunk a foot or two in the last decade and that the area around Wilmington has sunk nearly 20 feet in the last 30 years. The old Edison Company steam plant at Wilmington is now below sea level. This sinkage is thought to be due primarily to the withdrawal of fluids, oil, water, and gas from beneath the ground. It is of artificial origin rather than a result of natural deformation.

RESOURCES

The Los Angeles Basin, with some forty-six separate oil fields, is a prolific petroleum area. Total production since about 1880 exceeds 5 billion barrels, nearly half the total for all of California. Most of the oil is pumped from Lower Pliocene and Upper Miocene rocks, and most of the fields are on anticlines.

California has not produced much gold in recent years, but several years ago sand and gravel pits at Azusa were the major gold producers of the state. This gold was obtained as a byproduct of sand washing. It represented a placer accumulation with an original source in gold-quartz veins up the East Fork of San Gabriel River.

SPECIAL FEATURES

Palos Verdes Hills. Palos Verdes Hills rise like an island above the alluvial sea of Los Angeles Basin. Indeed, they were an island in the ocean less than a million years ago, and their bedrock basement is closely allied to that of Catalina. Structurally, the Palos Verdes Hills are a doubly plunging anticline uplifted along a fault on the northeast side (Figure 2-7). A small area of Catalina schist is exposed in the core of this anticline high on the northeast flank along the headwaters of George F. Canyon, between Crest Road and Palos Verdes Drive. A dozen smaller anticlines and synclines are exposed on the flanks of the main fold; so you will see beds inclined in just about all directions in different road cuts.

The most widely exposed rock unit is the Monterey Formation, a thinly bedded, brown-to-white shale which covers fully 90 per cent of the surface. Locally it is rich in diatoms, the skeletons of microscopic single-celled plants that float around in water. A quart of sea water can contain a million diatoms. Diatom skeletons are made of silica, and where richly concentrated they compose deposits of diatomite which is used in filters and adsorbers. Diatomite was formerly quarried in the Rolling Hills district of Palos Verdes Hills.

Overlying the Miocene rocks, mostly on lower slopes around the edges, are thin deposits of younger Pliocene and Pleistocene beds. Some of these younger accumulations are richly fossiliferous, furnishing a variety of sea shells in great abundance and a good state of preservation. The old Second Street locality in San Pedro has long been noted for its Pleistocene marine fossils.

Palos Verdes Hills are justly famous for many things, but one of their more infamous features is the landslide at Portuguese Bend. The following constitutes a recipe for disaster: Take a series of soft, thinly bedded, incompetent shale layers, incline them southward toward the sea, let the sea be constantly undercutting and steepening the slope, and add some lubrication in the form of altered clay-rich volcanic ash and water. The result—almost inevitably a landslide.

The Portuguese Bend slide had probably moved many times before settlement of the area, but in 1956 it slipped again destroying almost totally about 100 houses and seriously damaging at least 50 more at an estimated loss of 10 million dollars. The courts have allowed a claim of over 5 million dollars against the County of Los Angeles; so all county tax payers have a stake in the problem.

The pity is that this area had obviously undergone sliding in the past and renewed movement would have been a reasonable expectation. It is testimony to the inadequacy of our zoning practices that the area was not early designated as a park site for riding trails, picnic grounds, and open-space usage. The slide moves slowly, at most a few inches per day, so recreational uses could conceivably be compatible with its behavior.

A striking feature of Palos Verdes Hills is the marine terraces, thirteen in all, rising as a succession of stair-like steps from sea level to an elevation of 1300 feet (Photo 2-1). These are particularly prominent at the west end of the hills in the Lunada Bay area. The next to lowest terrace, about 140

Photo 2-1. Uplifted, wave-cut, marine terraces at west end of Palos Verdes Hills, viewed southeastward. (Photo by R. C. Frampton and J. S. Shelton, 4-9064B, 1953).

feet above sea level, is the widest and best preserved. Many of the vegetable and flower growing plots as well as houses, roads, and highways, such as Palos Verdes Drive, are on terraced surfaces. Higher terraces are largely obscured by erosion and by burial beneath sloping surfaces of later alluvial deposits.

These terraces demonstrate that much geological deformation occurs in episodes or steps. Each terrace represents a stable period of little or no deformation when the sea was cutting horizontally back into the land (Photo 2-2). Such stable periods were terminated by uplift which raised the wave-cut platform and the sea cliff at its inner edge, converting them to the stair-like step we call a terrace. Thus, Palos Verdes

Hills attained the present height of 1480 feet in at least thirteen separate episodes of uplift, probably more. Remember this when you look at the 8,000-foot scarp along the east face of the Sierra Nevada. This scarp wasn't created all at once, rather it represents a summation of countless uplifts, 5 feet here, 3 feet there, and 20 feet some other time.

Antecedent streams. Some of you have driven north or south on San Gabriel River Freeway (605) through the Whittier Gap or Narrows, perhaps countless times. Did you ever stop to wonder why the gap exists? If you did, you probably realized that the San Gabriel River which flows through the gap had something to do with it. But just imagine for the moment that you were the

Photo 2-2. Slide on face of Pacific Palisades blocking Pacific Coast Highway west of Santa Monica. (Photo by John S. Shelton, 987).

San Gabriel River flowing south from Azusa before there was a Whittier Narrows. What would you do then?

The proper question really is, Which was there first, the Whittier Hills or the San Gabriel River? In most parts of the country the preferred answer would be the hills. Rivers are mostly young, geologically speaking, and hills are usually older. In California, however, we just can't conform. Geological relationships suggest that the San Gabriel River was here first and that the

Whittier Hills were uplifted across its path. Uplift was slow, and the river was able to cut down and maintain its course. Thus, the Whittier Narrows were cut gradually as the hills rose. The river anteceded (came before) the hills, and in its course across the Whittier Hills it is an antecedent stream.

The Los Angeles Basin displays other examples of stream antecedence, such as Los Angeles River across the Elysian Park anticline near Riverside Drive and across the Dominguez Hills alongside the Long Beach Freeway, and Santa Ana River across the Santa Ana Mountains between Corona and Anaheim and across Newport Mesa near Costa Mesa. Antecedent streams are rare birds, and we have a handful right here at home. Be proud of them.

MOJAVE DESERT

People from the leafy greenness of better watered parts of the world can be less than enthusiastic about deserts. That's a shame, because desert country has a charm all its own. Most geologists love the desert since everything geological is so well exposed; no false eyelashes, no cosmetics, no fancy clothes, just pure plain naked geology with a good coat of tan (desert varnish). A little acquaintance with arid-region geology may help develop your taste for the desert, making travels therein more fun. If you already like the desert, geology can greatly increase your enjoyment of it.

The Mojave is neither one of the largest nor one of the harshest deserts, but geologically it must be one of the more varied. It has just about everything in the way of rocks, landscape, structures, and processes.

CHARACTER AND DIMENSIONS

A glance at Figure 2-8 shows two things about the Mojave Desert. It is the largest of our southern California provinces, and the wedge-shaped western end (Photo 2-3) looks like the bow of a ship sailing into the rest of California. This wedge is outlined by features related to two of California's largest fault zones, the San Andreas on the south and the Garlock on the north. The analogy with a ship sailing west would be reasonable except for one thing—the captain has the engines reversed. The right-lateral slip on the San Andreas and the left-lateral slip on the Garlock indicate the Mojave block is moving relatively east, not west.

For convenience we refer to the western Mojave as that part lying within this wedge and bounded on the east by the Mojave River and a line running northwest from Barstow to Red Rock Canyon (Figure 2-8). The eastern Mojave comprises the remainder and larger part, extending to the Nevada border and the Colorado River.

Topographically the eastern and western Mojave are distinct. The west consists of great expanses of gentle surface with isolated knobs, buttes, ridges, and local hilly areas. It is a striking landscape when shadowed and highlighted by low-angle morning or evening illumination. Under these conditions isolated knobs look like ancient castles guarding the land. The far western part is exceptionally flat, being an area of deep alluvial fill. Eastward the alluvium thins, and much of the gentle terrain is bedrock reduced to low relief by weathering and erosion. This western region also harbors several good sized dry lakes of which Rosamond, Rogers (Muroc), and Mirage are the best known.

Although smooth and low relative to bordering mountains, the western Mojave is hardly a lowland. Much of it is 2000 to 2500 feet above sea level, which is higher than the top of Palos Verdes Hills (1480 feet). The residual knobs and ridges, so called because they are erosional remnants

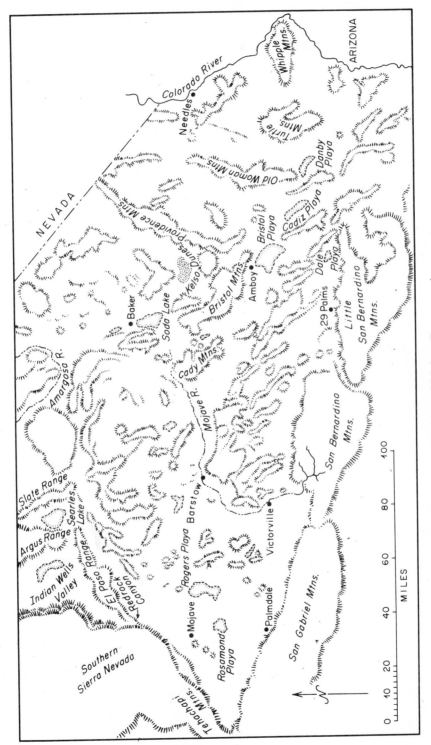

Figure 2-8. Map of Mojave Desert province.

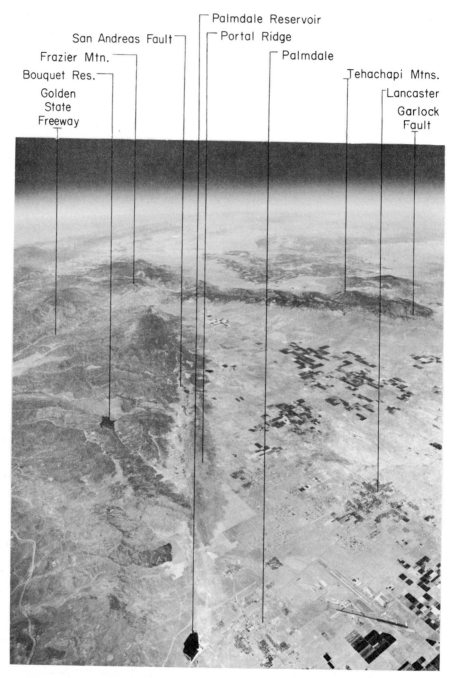

Photo 2-3. Very high-altitude oblique view of western Mojave Desert, looking northwest along trace of San Andreas Fault. (Photo taken by U. S. Air Force for U. S. Geological Survey, 064R-140).

of once much larger masses, are still higher by 1000 to 2000 feet. Soledad Mountain just west of Mojave town is 4183 feet above sea level.

The eastern Mojave also has its share of gentle topography but largely in the form of basins and open valleys between mountainous masses. In the southern part, these mountains and valleys have a northwest alignment, like a herd of caterpillars heading for lunch. In the northern half, the caterpillars are either well fed or haven't heard the dinner bell. Some are headed northeast, some west, some northwest, and some south—confusion reigns.

In addition, two large troughs extending east-southeastward are out of phase with any pattern. One starts from Barstow and is followed about two-thirds its length by U. S. Highway 66 and the Santa Fe Railroad. A similar somewhat smaller trough runs parallel from Victorville to a *cul de sac* at Dale Dry Lake, 20 miles east of Twenty Nine Palms (Figure 2-8). Apple and Lucerne valleys lie in this trough which has no major road traversing its full length. Some people regard these troughs as possibly the product of streams formerly draining to the Colorado River.

The eastern Mojave has its greatest relief to the east. There, some of its valleys are only 600-800 feet above sea level, and many higher peaks and ridges attain elevations of 4500 to nearly 7500 feet. The north face of the Providence Mountains, as viewed from Kelso Dunes in the late afternoon, is an impressive mountain scarp rising 5000 feet above Kelso Valley.

ROCKS

If someone asked which southern California province would supply the greatest variety of rocks for an avid collector, the answer would surely be, "Mojave Desert." The menu offered equals that of a first-class cafeteria. There are old-old rocks (early Precambrian), old rocks (late Precambrian), medium-old rocks (Paleozoic), not-so-old rocks (Mesozoic), and young rocks (Cenozoic). Each class, igneous, metamorphic, and sedimentary is represented in delightful variety.

The old-old rocks (early Precambrian) are mostly gneisses and associated metamorphic types. However, not all metamorphic rocks in the Mojave are necessarily very old. Some of the not-so-old rocks (Mesozoic) are locally highly metamorphosed. However, early Precambrian rocks are invariably strongly metamorphosed. They have been around too long to have escaped severe squeezing, heating, and recrystallization.

Some of the highly metamorphosed rocks of this region are known to be early Precambrian for another reason. They are overlaid by 7000 feet of only slightly metamorphosed conglomerate, sandstone, shale, and carbonate beds of late Precambrian age. This occurs in the Kingston Range just outside of the northeast boundary of the Mojave Desert as defined here (*see* Figure 2-13). These unfossiliferous late Precambrian sedimentary rocks are overlaid in turn by 20,000 feet of fossiliferous medium-old (Paleozoic) sedimentary rocks.

Highly metamorphosed exposures of late Precambrian, Paleozoic, and Mesozoic rocks seen in other parts of the Mojave commonly occur as large inclusions within igneous bodies. For example, the marble mined for cement near Victorville is of the same age as unmetamorphosed limestone in the Paleozoic sequence of the far-eastern Mojave.

Mesozoic metamorphics are exposed mostly in the west-central part of the eastern Mojave and are notable for their high content of volcanic material. This was only one of several intervals of volcanism in the region. However, the most abundant and widespread Mesozoic rocks are coarse-grained granitic intrusives. A generalized geological map of Mojave Desert shows that about 50 per cent of the surface is alluvium. Of the remainder, at least 40 per cent by conservative estimate is granitic rock, largely of Mesozoic age. Some of these Mesozoic granitic intrusives have been dated as 150 m.y. old.

Granitic rock is normally regarded as durable, and it is in most places. However, in the desert, granite disintegrates so readily that it wears down more rapidly than most metamorphic, volcanic, and hard sedimentary rocks. Thus, some of the lower smoother parts of the Mojave are underlaid by coarse-grained granitic intrusive bodies.

Following the Mesozoic igneous activity, the Mojave area was uplifted and stood as a relatively high land mass shedding rock debris to other areas. A thickness of rock amounting to roughly 20,000 to 25,000 feet was removed at this time. By the middle of the young-rock period (Cenozoic) the Mojave region was topographically subdued. Then in about mid-Cenozoic time, warping and faulting created local basins into which rock detritus of local origins was carried. Deposits in some of these basins attained thicknesses of many thousands of feet. Concurrent volcanism created thick piles of lava and contributed large quantities of fragmental volcanic detrius to the fillings in the basins. Volcanism has continued at intervals practically to the present.

You may wonder why so many of the rocks in the desert are dark brown or black. Actually, they are of all colors, but a thin coating rich in iron and manganese known as *desert varnish* makes their surfaces dark. Varnish forms in different degree on different rocks and is strongly controlled by the environment. Deserts provide optimal conditions for its development and preservation.

STRUCTURES

Let's return to our herd of caterpillars in the south. They appear to be crawling northwestward because the country, like a loaf of bread, is sliced by over twenty major and many minor northwest trending faults. Some of you may recognize a familiar name or two in this list: Mirage Valley, Blake Ranch, Helendale, Muroc, Old Woman Springs, Lenwood, Lockhart, Harper Lake, Camp Rock, Copper Mountain, Calico, Blackwater, Mesquite, Pisgah, Ludlow, and Bullion. These are but some of the faults. The same pattern extends into the western Mojave, but it is less clearly expressed in the subdued topography of that region. Scarps breaking young alluvial surfaces along some faults indicate relatively recent displacements. Signs of right-lateral offsets along them suggest that these faults may be related to the San Andreas family.

In the north the caterpillars are confused, crawling in all directions. As you surmise, this is because the faults in that area are inconsistent in trend. Some bearing northwestward are extensions from the southern group, but a good many trend northeast, east-west, and even due north. Garlock fault bounds the Mojave on the north, and it has here swung into an east-west course (Photo 3-10). With its left lateral displacement, it represents a very different stress arrangement than the San Andreas. This may be a source of much of the confusion.

A north-south line through Baker (Figure 2-8) roughly marks a boundary in the eastern Mojave between two types of country. Keep your eyes open if you drive to

Las Vegas, and you can sense this. To the west are irregular mountain ridges and masses composed largely of metamorphic, intrusive igneous, and volcanic rocks. To the east are higher, better defined, more linear mountain masses which display thick sections of well-layered Paleozoic sedimentary rocks. The separation between these regions is roughly the trace of the Furnace Creek-Death Valley fault zone (Figure 1-1). Farther north this is a very prominent structure. Here in the Mojave it is less clearly defined, but nonetheless it may be one of the more significant structures of the region.

In the vicinity of the Furnace Creek-Death Valley fault zone, for example in the Shadow Mountains a little northeast of Baker, are low-angle (nearly horizontal) thrust-fault plates. Here early Precambrian rocks have been thrust over Cenozoic deposits. Farther east, thrust faults become larger and more numerous until in southern Nevada, northwest of Las Vegas, they are truly spectacular.

RESOURCES

Since the Mojave Desert has a varied geology, one might expect a variety of natural resources. Indeed, most of the common metallic elements have been sought and discovered, with significant production of some: gold and silver at Mojave; gold, silver, and tungsten at Johannesburg, Randsburg, and Atolia; silver at Calico. Additionally, iron, copper, lead, zinc, molybdenum, antimony, tin, uranium, and thorium have been produced in smaller amounts. Currently, some of the unusual products of the area are rare-earth elements such as cerium, lanthanum, and neodymium from the Mountain Pass area just north of the Las Vegas highway near Clark Mountain.

At present the more valuable products are nonmetallic materials, of which borax is the most notable. About midway between Mojave and Barstow, near Boron, is the greatest borax producing area the world has ever known. Here, layers of borax minerals occur in lake beds deposited in a Cenozoic basin of terrestrial sedimentation. The deposits are now efficiently mined in a huge open pit. The other major non-metallic product is cement which comes largely from marble bodies in the Victorville area, from Cushenbury Canyon north of Lucerne Valley, and from the Tehachapi Mountains east of Mojave. Volcanic cinders are mined for light-weight aggregate in several places.

SPECIAL FEATURES

Dry Lakes. Mojave Desert is an area of interior drainage, which means that it does not drain to the sea. The Mojave is the largest river of the province, and if it had enough water it would flow all the way from the San Bernardino Mountains to Death Valley. In fact it did just that, through a string of lakes within the last 15,000 years.

These are some reasons for thinking that at earlier, but still geologically young, times the Mojave River, with possibly a major tributary from the Death Valley country, flowed southeastward through Bristol, Cadiz, Danby, and Chuckwalla valleys (and lakes) to the Colorado River. Related species of fish in some of the present-day landlocked desert basins and features of the landscape support this interpretation of an integrated network of through-flowing streams. Now the drainage has become disintegrated, that is, it has been broken into unconnected bits and pieces. Climatic dessication, volcanic activity, faulting, warping, and building of alluvial divides (mostly fans) have accomplished this dis-integration. In the process many separate basins were created, the floors of which are now occupied by dry lakes—at least they are dry until flooded by occasional local heavy rains.

The floors of these lakes are called *playas*. You should make their acquaint-ance. A playa is one of the flattest natural features on land. One-half inch of water usually suffices to cover many square miles of a playa. If a wind comes up, all the wa-ter may blow to one end. Playas are fun to drive or camp on, but keep off when wet. Even if you don't get stuck, which would serve you right, you mess them up horribly with deep wheel tracks.

Geologists sometimes speak of "wet" playas. This reference is to playas with underground water at such shallow depths, 10-20 feet, that moisture can move up small capillary passages to the surface. As it evaporates, this capillary water deposits salts in the soil which fluff up the ground like a good pie crust. The result is an un-even puffy surface that is poor for driving and into which you sink 2 or 3 inches on foot. Soda Lake south of Baker is a "wet" playa in places, as is Lucerne Dry Lake and many others.

Recent volcanoes. Volcanoes are fasci-nating as long as they aren't too hot, too ac-tive, or too close. The Mojave has some beauties. They are perfectly safe even though some were active within the last few thousand years. The most accessible is Pis-gah, 35 miles east of Barstow alongside U. S. Highway 66 to Needles. Pisgah vol-cano is a cinder cone about 250 feet high surrounded by a succession of lava flows, some of them very young. *Cinders* are simply small broken chunks of highly por-ous volcanic rock. The cinders at Pisgah are mined for light-weight aggregate, and

the mining company usually keeps the road to the cone blocked part way in. That, how-ever, need not prevent you from going as far as you can, and then walking around on the surface of the lava flows or climbing the cone to look down into its central crater.

Amboy Crater and associated lava flows just west of Amboy are about as fresh but a little harder to get to. If you really like volcanoes, take the Kel-Baker road south-east from Baker. On its way to Kelso, it passes alongside a volcanic field with some twenty-six separate cones and a number of lava flows.

The Barstow Formation. Of outstand-ing interest among land-laid Cenozoic de-posits in the Mojave is the Barstow Forma-tion, which consists of about 5000 feet of sandstone, siltstone, shale, and fine volcanic debris. These beds are well exposed in Rainbow Basin, 10 miles north of Barstow. Interest in the Barstow beds comes pri-marily from remains of vertebrate mammals living in the area some 15-20 m.y. ago, in-cluding dog-like bears, mastodons, a variety of large and small horses, camels, prong-horn antelope, pigs, dogs and hyena-like dogs, saber-toothed cats, deer, and a variety of small rodents. The country was then more moist, and the abundance of grazing animals suggests extensive grass lands. There were also some palm trees. Rainbow Basin is now protected as a State park.

SALTON TROUGH

The heading properly brings to mind the Salton Sea and Imperial Valley. However, viewed from the southeast, one sees that the Salton Trough is basically a landward con-tinuation of the Gulf of California extend-ing northwestward all the way to San Gor-gonio Pass. It thus includes Coachella Valley and Palm Springs. In addition, the

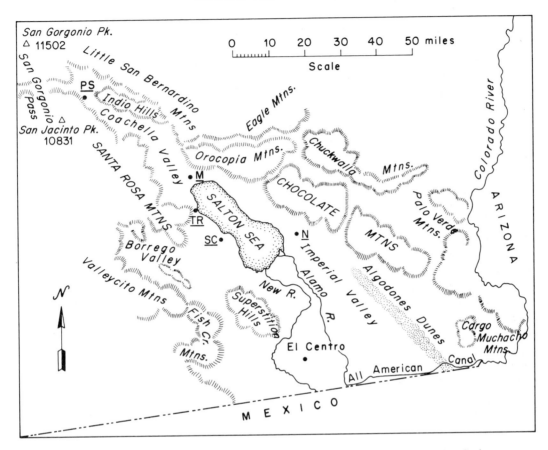

Figure 2-9. Map of Salton Trough province; M—Mecca, N—Niland, PS—Palm Springs, SC—Salton City, TR—Travertine Rock.

northeastern boundary is drawn to encompass mountain ranges that don't fit comfortably into either the Transverse Ranges or Basin Range provinces (Figure 2-9). These ranges are the Orocopia, Chocolate, Chuckwalla, Palo Verde, and Cargo Muchacho, intriguing names and geologically interesting areas.

Salton Trough and its immediate environs provide a geological menu of considerable variety. Items for consideration include large faults slicing across the area, recent volcanic knobs at the southeast end

of Salton Sea, hot brine wells, sand dunes, Colorado River flooding, and the past history of the Salton Sea.

CHARACTER AND DIMENSIONS

Salton Trough embraces the largest area of dry land below sea level in the Western Hemisphere, something over 2000 square miles. Death Valley is a little deeper, -282 feet compared to -273 feet, but the part of Death Valley below sea level is much smaller.

Salton Trough extends 140 miles north-westward from the head of the Gulf of California to San Gorgonio Pass. It is only a few miles wide at the northwest end but attains a maximum width of roughly 70 miles at the Mexican border. Geologically this is a structural feature of impressive vertical dimensions. East of Mecca, on the northeast margin, the buried bedrock basement abruptly rises 12,000 feet. Under Imperial Valley the depth of unconsolidated fill exceeds 20,000 feet.

The trough is bordered by rugged mountains on both sides, those to the west being higher and more massive. The western margin southward from the end of the Santa Rosa Mountains is irregular owing to major embayments, such as Clark and Borrego valleys, and forelying hills, such as the Superstitions. This is the result of complex geological structures.

ROCKS

The bordering mountain ranges consist largely of metamorphic and igneous rocks, some old (Precambrian) and some not so old (Mesozoic). The units recognized include the Chuckwalla gneiss, Orocopia schist, and intrusive granitic rocks related to the southern California batholith of the Peninsular Ranges.

The trough itself is filled with young (Cenozoic) sedimentary deposits, largely of land-laid origin but partly marine. These land-laid beds change rapidly from place to place and coarse gravels commonly intermix with layers of finer sandstone, siltstone, and mudstone. Locally some volcanic rocks are included. The oldest Cenozoic beds probably do not exceed 25 m.y. in age. These deposits are thickest in the southern part of the trough where a well 12,000 feet deep failed to penetrate the section, and geophysical explorations indicate about 20,000 feet of soft sediments above the basement.

Within the Cenozoic units the Imperial Formation is unusual. It is composed mostly of brownish sandstone and mudstone, locally rich in marine fossils (snails, clams, oysters, and corals). The Imperial beds are preserved in isolated patches from the Mexican border almost to San Gorgonio Pass. The fossils are not normal Pacific-Coast types. They are more tropical and apparently migrated up the Gulf of California. The Imperial Formation is estimated to be 10-15 m.y. old.

In Fish Creek Mountains along the west side of lower Imperial Valley, a deposit of lake beds rich in gypsum underlies the Imperial Formation. It is extensively mined and processed into plaster and plaster-board products at Plaster City. Other unusual rocks are found in the volcanic knobs at the southeast end of Salton Sea. Here is *obsidian* (natural volcanic glass) and *pumice* (frothy volcanic glass), so light it floats on water. The floor of Imperial Valley is veneered by fine grained lake deposits, locally rich in snail and clam shells. These were laid down when expanded stages of the Salton Sea formerly covered much of the valley.

STRUCTURE

Southern California's three big right-lateral fault zones, the San Andreas, San Jacinto, and Elsinore, traverse this province. In fact, there are so many northwest trending faults in the area that it's hard to know to which system individual fractures belong. This is partly due to the fact that the San Jacinto fault particularly, and the

Elsinore to some degree, splay out into separate branches as they approach Imperial Valley.

Representatives of the San Andreas system can be traced along the northeast side of Salton Trough to the Algodones Dunes (Sand Hills). This dune belt is aligned along a fault which is probably the southern continuation of the San Andreas. The San Jacinto separates into a complex of faults in Borrego Desert and southward. Among these fractures, the Imperial fault is particularly noteworthy as it has generated two sizeable earthquakes within modern historical times, 1915 and 1940. It can be traced for many miles by a 10-15 foot scarp extending northwest about 3.5 miles east of the town of Imperial. If one looks carefully at the alignment of powerline poles, highways, railroads, the All American Canal, and even the Mexican border where crossed by the Imperial fault, they all show the effects of right-lateral displacement. The border was offset nearly 15 feet in the 1940 earthquake. Branches of the Elsinore fault help define the western border of Salton Trough and continue into Mexico.

Imperial Valley faults are currently experiencing slow continuous slippage. This produces cracks in highways where they cross the fault trace. Highly precise resurveys of Coast and Geodetic markers demonstrate current differential slip of opposing blocks along the Imperial Fault of 1-2 inches a year.

Folds are displayed along the margins of Salton Trough where the young (Cenozoic) sedimentary filling has been uplifted. They can be seen off the southeast end of Santa Rosa Mountains and south into the Superstition Hills. Folds are also found along the northeast margin of the trough in the Mecca and Indio hills. A ride up Painted Canyon east of Mecca provides a good view of severely deformed Cenozoic beds. By watching the inclination of bedding you can recognize several folds. One or two of the anticlines have old metamorphic rocks squeezed into their cores.

Geologists have recently concluded that the Gulf of California was created within the last few million years by ripping Baja California away from mainland Mexico and moving it to the west-northwest. The force is thought to have been supplied by the same sort of deep convection currents within the earth that cause spreading of the sea floor outward from mid-oceanic ridges. Since Salton Trough is simply an extension of the Gulf of California, it too has presumably been formed, or is being formed, by the same process, but it is in a less advanced stage of development. If the process continues, San Diego may eventually find itself separated from Arizona by an arm of the sea.

SPECIAL FEATURES

Salton Sea. Between 1900 and 1905 the floor of Salton Trough was dry, although in the 1880's and again in 1891 overflows from the Colorado River had created lakes a few feet deep in its lowest part. Starting in 1905 the main flow of the Colorado River was accidentally diverted by way of irrigation canals into the Salton Trough. This was the beginning of the present Salton Sea. Only the herculean efforts of the Imperial Irrigation District and the Southern Pacific Company restored the river to its normal channel by 1907. By then flood waters flowing down the channels of the Alamo and New rivers had created a lake nearly 80 feet deep, with a water level and shoreline at −195 feet, covering over 400 square miles. Once inflow was shut off, the lake began to shrink owing to intense evaporation, and by 1925 it had become stabilized at –250 feet. The lake fluctuated around that level until 1935 when it began to rise again. It has risen steadily ever since to the 1970 level of –230 feet. Completion of the All American Canal, thereby introducing more water into Imperial and Coachella valleys, was the cause of this last rise.

About 6 feet of water are evaporated from the surface of Salton Sea each year. As its size increases so does the area from which evaporation occurs. Eventually a balance will be struck between inflowing water and evaporation loss. Calculations suggest that this will happen before 1980 and at some level below –220 feet. Accordingly all lands bordering Salton Sea below –220 feet have been withdrawn from public occupation in order to prevent economic loss. The rising level has already made islands out of some of the volcanic knobs at the southeast end of Salton Sea and has submerged the former hot springs, steam jets, mud pots, and mud volcanoes of the Niland area.

Since human beings accidentally created a good sized lake in 1905-1907, the question arises, "Might not nature have done the same thing, and perhaps on a larger scale, at some earlier times?" Around the edges of Salton Trough at about 40 feet above sea level are prominent shoreline features of a lake well over 300 feet deep which occupied the trough about 300 years ago. This water body goes by the names Lake Cahuilla or Lake Le Conte. It appears to have been held at the +40 foot level by the crest of the Colorado River delta separating Salton Trough from the Gulf of California.

Along the highway on the east side of Salton Sea a prominent Lake Cahuilla beach ridge of gravel can be seen back of Hot Mineral Spa. Along the west side a superb Lake Cahuilla shoreline is cut into the mountain front in the vicinity of Travertine Rock (Photo 2-4). Rocks above the shoreline appear light because of wave cutting which produced a cliff and destroyed the desert varnish still preserved at higher levels. Rocks below the shoreline are coated with dark travertine. The *travertine* (a porous, irregular deposit of calcium carbonate on rock surfaces) of Travertine Rock alongside the highway was deposited on rocks below the surface level of Lake Cahuilla. The encrustation is as much as 30 inches thick.

There are scattered remnants of shorelines of still much older and larger lakes around the margins of the trough. Some of these bodies date back thousands of years, and judging from the shell life living along their shores, one or two had connections with the gulf. This means that the oceans either stood higher then or that the shoreline features have subsequently been uplifted. The latter seems more likely.

Sand dunes. Many of you have already seen a lot of the Algodones Dunes (Sand Hills) along the southeast edge of Salton Trough without leaving your home. They are favored by movie and TV companies filming episodes purporting to occur in the Sahara. These dunes are also much used by dune-buggy enthusiasts.

Algodones Dunes comprise a chain which is 45 miles long in a north-northwesterly direction and 4-8 miles wide. Sand within this belt attains a maximum thickness of nearly 400 feet. One of the impressive features of this chain is a series of subequally spaced intradune flats, seen particularly well in the southern third of the belt from the air (Photo 2-5). These flats are egg shaped areas inclosed by dunes from which sand has been removed by the wind down to the un-

Photo 2-4. Horizontal shoreline markings formed by expanded Salton Sea (Lake Cahuilla phase) at base of Santa Rosa Mountains near Valerie Jean. (Photo by John S. Shelton, 3600).

Photo 2-5. Intradune flats in southern part of Algodones Dunes, viewed northward. U. S. Highway 80 and All American Canal traverse second flat. Flat in foreground and surrounding dunes much used in motion picture films. (Photo by John S. Shelton, 1816).

derlying base of gravel. The All American Canal and U. S. Highway 80 cross the dune chain by means of a small intradune flat. The flats are dynamic, moving south-southeastward at 6-12 inches per year. This movement will eventually bring a huge dune mass right up to the All American Canal. However, it looks as though that unhappy day is about 400-500 years in the future. By that time the canal will have paid for itself several times over.

On the west side of Salton Sea in the vicinity of the old Salton Sea Test Base, 7 miles south of Salton City, is a field of barchan dunes (Photo 2-6). Barchans are individual dunes of symmetrical crescentic shape.   They develop best in areas where the wind tends to blow mostly from one direction and where the sand supply is not too abundant. The crescents are open downwind. Little barchans move more rapidly than large ones, so sooner or later they catch up with larger dunes which gobble them up, making the large barchan even bigger and slower.

Hot brine wells. Interest in geothermal power, that is, power generated by steam from natural sources, brought attention to the area of the youthful volcanic knobs, hot springs, and mud pots at the southeast corner of Salton Sea near Niland. At one time carbon dioxide was produced commercially from shallow wells in this area.

In the 1960's a number of wells were drilled here to depths of 5000-8000 feet. Several of them have yielded hot brine at temperatures of 400-600°F. When brought to the surface much of the water in the brine

Photo 2-6.   Barchan dunes west of Salton Sea south of Salton City, viewed north-northwest. (Photo by John S. Shelton, 1807).

flashes to steam. The steam resources of the area have been estimated as adequate to provide electric power for a population of 4 million people for about 8000 years. This sounds great, but there are some problems. First, the brine is so corrosive that it is hard to handle. Second, some means is required for disposing of the brines without contaminating the ground and surface water. This can perhaps ultimately be accomplished by injecting them back into the ground by way of deep wells.

The brine contains chemically useful materials, particularly potassium, valuable as a fertilizer, and lithium, used in a variety of chemical compounds. The problem is again one of proper handling and extraction. Pilot plant operations are underway, and eventually the hot-brine wells of Imperial Valley will probably yield both power and chemicals. Besides the elements noted, the brine contains gold, silver, copper, nickel, chromium, lead, zinc, iron, manganese, sodium, calcium, barium, strontium, boron, chlorine, carbon, and goodness knows what else. You can see it is a real witch's brew.

SOUTHERN COAST RANGES

We treat the southern Coast Ranges briefly, but only because they are peripheral to our geographical focus. Southern Coast-Ranges country is delightful, and the geology is good. Take the western sector of the Transverse Ranges with its thick sedimentary section and many faults and folds, rotate it so the grain runs northwest, and there you have the basic characteristics of the southern Coast Ranges. Add a rugged coastline and some of the loveliest coastal dunes of the Pacific shore, south of Pismo (Photo 2-7), and you begin to catch the

Photo 2-7. Coastal dunes near Pismo. Onshore winds carry sand inland from beach. (Photo by John S. Shelton, 4393).

essence of the country. Morro Rock is a volcanic plug, the last one in a line extending northwest from San Luis Obispo (Figure 2-10).

The sedimentary section in the southern Coast Ranges is thick, and every geological epoch from Cretaceous (*see* Appendix A) up is represented by layers of sandstone, conglomerate, or shale. Deformation has occurred intermittently over many tens of millions of years, and the oldest beds are the most severely folded and faulted. How-

ever, movement has continued right up to the present; so even young beds display some structure.

The region boasts a host of faults, two of which, both bearing northwesterly, merit particular notice. One is our old friend, the San Andreas, and the other is a parallel fault, about 30 miles to the west, the Nacimiento (Figure 1-1). No place is better than Carrizo Plain, near the eastern edge of the southern Coast Ranges (Figure 2-10), in which to see the marks of the San An-

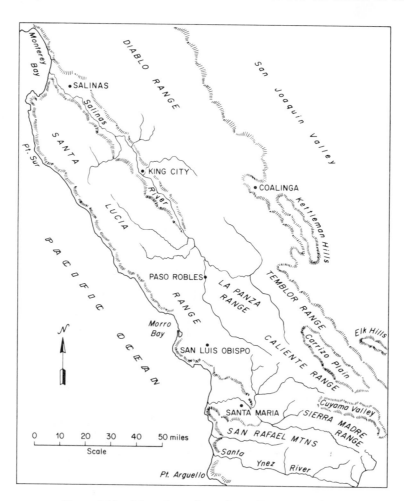

Figure 2-10. Map of southern Coast Ranges province.

dreas upon the land. Commuter flights between San Francisco and Los Angeles not infrequently travel right along the San Andreas in this area; so keep your eyes peeled. Owing to gentle relief, scanty vegetation, and the transverse crossing of drainage courses by the fault, the evidence of displacement is spectacular (Photo 2-8). The fault makes a fresh scar across the land. Within the fault zone are small depressions, marshes, and wet spots, called *sags,* or *sag ponds* when wet. They mark places where a slice within the zone has sunk. Uplifted slices compose fault ridges. Scarps and elongated narrow troughs are abundant. Clearly evident is the lateral offset of stream courses, showing right-lateral displacement along the most recent line of movement.

The Nacimiento fault is regarded as relatively old, but not necessarily dead. Like old scars on the human body, old faults on Mother Earth are partly obscured by healing. The Nacimiento is a complex zone which extends from Cuyama Valley northwest at least to Point Sur, and it probably continues on offshore into the submerged continental borderland. It separates two distinct types of basement rocks. To the northeast are granitic rocks, perhaps 90 m.y. old, intruding an older metamorphic complex of gneiss, schists, quartzite, and marble—the Sur Series. To the southwest is a wholly different series of peculiar, metamorphosed, dark sandstone, shale, conglomerate, volcanic, chert, and limestone layers known as the Franciscan Series. Associated are some iron-magnesium rich intrusive masses, altered to a slippery soapstone, known as *serpentine.* Remember that the Newport-Inglewood fault zone in the Los

Photo 2-8. Trace of most recent break along San Andreas fault zone in Carrizo Plain, looking southeast. (Photo by John S. Shelton, 510).

Angeles basin separated two distinct types of basement rocks. The Catalina schist of the southwestern block there may be related to the Franciscan Series. Total displacement on the Nacimiento is not known, but it is judged to have been considerable. Some has been lateral and some, particularly of younger date, has been vertical.

Before leaving the Coast Ranges, the diatomite deposits of the Lompoc area merit mention. This is a truly notable accumulation of these little silica skeletons of single-celled marine plants. The skeletons are extremely irregular; hence they have a very large surface area for their size. It has been said that a pound of diatomite has a surface area equivalent to that of three to eight football fields. This coupled with the fact that silica is a stable, chemically unreactive material makes diatomite highly prized as a filtering and adsorbing material. It is extensively mined in the Lompoc area for commercial use.

THE GREAT VALLEY

The Great Valley of California lying between the lofty Sierra Nevada and the lower Coast Ranges is a true heartland. It actually consists of two valleys, the Sacramento in the north and San Joaquin in the south, each drained by an axial stream of corresponding name. Although these two sectors are part of a coherent feature, they display some differences in geology and resources. For example, this province has produced 10 billion dollars worth of oil and gas. Only gas is obtained in the Sacramento sector, while both are produced in the San Joaquin Valley. This contrast arises primarily from the differences in the nature of the underlying sedimentary deposits.

Topographically the Great Valley is mostly a smooth alluvial plain a few hundred feet above sea level. Geologically it is a structural trough with a northwest length of 450 miles and a width of 50 miles. In terms of basement rock it is strongly asymmetrical, shallow on the east and deepest near the western margin. This trough is filled by a wedge of sedimentary materials accumulated over an interval of nearly 150 million years. More than 60,000 feet of deposits have been laid down over the present valley site during this period.

In this booklet we are concerned only with the San Joaquin Valley, and specifically with that part south of Fresno (Figure 2-11). The two largest streams debouching into this section, the Kings and Kern rivers, have built huge alluvial fans of such gentle slope that you cross them almost without recognition. These fans block off parts of the valley creating shallow basins, Buena Vista on the south and Tulare on the north (Figure 2-11). Lakes and marshes occupied these basins until river waters were diverted into irrigation ditches and the land was drained for cultivation, much to the chagrin of duck hunters.

In high-water years Buena Vista Lake occasionally overflowed north to Tulare Lake, but the latter apparently seldom, if ever, topped the alluvial divide into San Joaquin River. These lakes and the north-flowing San Joaquin River hug the west side of the valley. They have been pushed there by large deposits of rock debris swept down by powerful Sierra Nevada streams only feebly opposed by small creeks from the relatively dry Coast Ranges.

Although the Great Valley has been the site of sedimentary deposition for nearly 150 million years, it assumed its trough-like form only during the last 60 million years. This coincided with initial uplift of the Coast Ranges. Westward tilting of the Sierra Nevada-San Joaquin block subsequently deepened the San Joaquin sector making it the principal site of marine deposition in the last 20-25 m.y. The scene must have been striking when a great inland sea filled the San Joaquin Valley and lapped against the west base of the Sierra Nevada or dashed its waves on a narrow coastal plain at its foot. Eventually this inland sea filled with sediment, and marine deposition was succeeded by terrestrial accumulation. The seas lingered longest at the southern end of San Joaquin Valley.

Although its present alluvial floor is nearly featureless, the valley has a good many folds and faults. Some of these structures are buried, but others are exposed along the west margin from Coalinga southward. Here a succession of outlying hills marks the site of a string of anticlines of which Kettleman Hills is the most famous. Oil and gas have accumulated in many of these anticlines. The folds are usually broken by faults, and one near McKittrick leaked oil to the surface creating tar pools

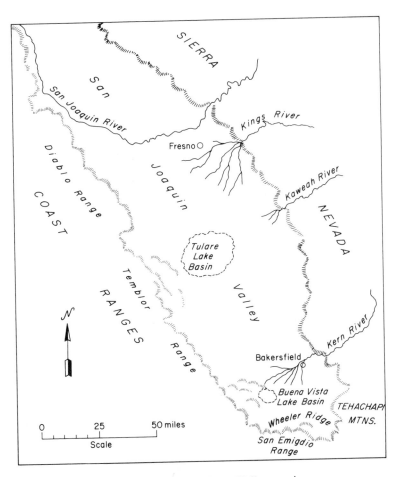

Figure 2-11. Map of Great Valley province.

in which animals and birds were trapped, just as at Rancho La Brea. The McKittrick fauna was not as rich and varied, but it was still impressive, including such animals as saber-toothed tigers, big lions, lynx, dogs, the big dire wolf, skunks, bears, ground sloths, camels, horses, bison, deer, antelope, elephants, mastodons, and lots of rodents. The abundant birds included grebes, herons, bitterns, storks, ibises, swans, ducks, geese, vultures, kites, hawks, falcons, partridge, quail, and cranes, among others.

Wheeler Ridge (Photo 2-9), a striking anticlinal fold of recent date, is seen just west of the Golden State Freeway as it leaves the foot of Grapevine Grade northbound. At the southern end of San Joaquin Valley the structures take on a nearly east-west trend. The folding of Wheeler Ridge is so recent that relatively unconsolidated sands and gravels are involved. These de-posits are currently being worked for sand and aggregate in a large pit on the north side of the ridge near its eastern end.

Wheeler Ridge anticline is asymmetrical, steepest on the north side, with a thrust fault along its north base. White Wolf Fault, which caused the destructive 1952 Arvin-Tehachapi earthquake, as traced west-southwest passes under Wheeler Ridge. The *epicenter* of the earthquake (spot on the ground beneath which slip first occurred) was just south of the ridge.

Wheeler Ridge is best seen traveling south on Golden State Freeway from Bakersfield, but northbound travelers can look back to view it in profile. Such a view shows that the crest is notched by several prominent topographic saddles or gaps. These were cut across the growing anticlinal fold by streams flowing north out of the San Emigdio Range to the south. Three such gaps can be seen near the east end of the

Photo 2-9. Wheeler Ridge west of Golden State Freeway beyond foot of Grapevine Grade. A recently formed anticlinal ridge with a wind gap, left center, and water gap near right (east) end, as viewed north. (Photo by John S. Shelton, 496).

ridge. The largest is no longer occupied by a stream, because the stream could not cut down fast enough to keep pace with the uplifting. It was turned aside. This abandoned notch is called a *wind gap.* Daniel Boone went to Kentucky through Cumberland Gap, a wind gap produced in a different way in the Appalachian Mountains. Our wind gap is used by the California Aqueduct as a means of crossing Wheeler Ridge. If the day is clear, you should be able to see the four large pipes rising into the gap from the pumping station at its north base.

SOUTHERN SIERRA NEVADA

Fine furniture, rugs, and drapes tastefully arranged can make a home attractive, but it helps to start with a good house. The Sierra Nevada are beautifully decorated with streams, lakes, meadows, and forests, and the air conditioning is tops. All this would not be nearly so attractive were it not associated with a spectacular bedrock landscape. This, California's largest mountain range, is an asymmetrical block, rising steeply from a major zone of faults at its east base (Photo 2-10) and sloping gently westward. Great masses of relatively homogeneous, coarse-grained granitic rock and deep sculpturing by running water and flowing ice are the elements giving the Sierras their special character. We are here concerned only with the south part and principally with the eastern side. A more complete treatment of the Sierras, specifically of the Yosemite-Tioga region, will be found in John Harbaugh's companion booklet, *Geology Field Guide to Northern California.*

Photo 2-10. East-face scarp of Sierra Nevada west of Lone Pine. Alabama Hills in foreground, Whitney Portal road on scarp face to right, Lone Pine Peak left center.

CHARACTER AND DIMENSIONS

The Sierras are over 400 miles long and as much as 80 miles across. Other ranges are higher, but not in conterminous United States. The difference in elevation between the highest and lowest points within the range is roughly 14,000 feet. The topographic asymmetry of the Sierras is extreme. The crest, in places, lies within 4-6 miles of the eastern base and 60-70 miles from the western foot (Figure 2-12). Consequently, streams flowing eastward are short and steep while those flowing west are 10 times longer, considerably larger, and traverse a gentler slope. Incidentally, the zone of maximum precipitation in the Sierras is far down the western flank, at elevations between 7000 and 8000 feet, not at the crest.

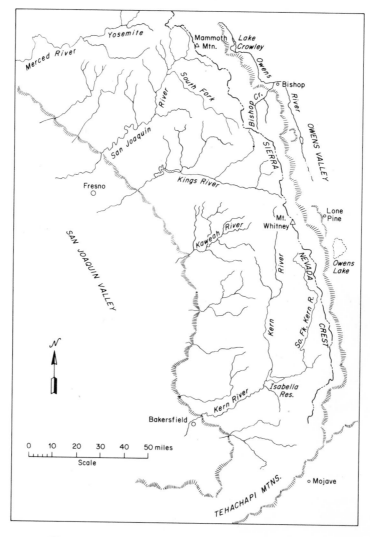

Figure 2-12. Map of southern Sierra Nevada province.

A few Sierra streams flow north or south roughly parallel to the range axis. An example is Kern River (Photo 2-11) which has eroded its valley along the fractured zone of a north-south fault. The upper part of the Middle Fork of San Joaquin River, in the Devil's Postpile region, flows south along a line determined by banded structures in metamorphic rocks. The direction of flow of Sierra streams is important for this reason. The range has been uplifted step by step, growing higher in the eastern part and tilting ever more steeply westward. Imagine that you are a stream flowing down the long western slope. Each uplift of the range steepens your bed giving you greater energy to cut deeper. Now imagine that you are a Sierra stream flowing north or south. You experience no increase of slope because of uplift, and furthermore, if you are tributary to a west flowing stream you will be left

"hanging" as the main stream cuts down. This is but one of several possible causes for the many hanging tributary streams in the Sierras with waterfalls at their mouths.

High up in parts of the Sierras are extensive areas of subdued upland terrain (Photo 3-25) across which travel is easy. The weathering characteristics of homogeneous granitic rocks have a lot to do with the smoothness of these upland flats. However, in places, remnants of such subdued surfaces make prominent benches along the sides of deep canyons (Photo 2-11) suggesting that at some earlier time the streams flowed in a wide, open valley before cutting down into their present narrow canyons. Thus, at least some of these features and areas of subdued relief are judged to be the product of extended periods of weathering and erosion occurring between the intermittent uplifts of the range.

Photo 2-11. Low-altitude oblique view south down Kern Canyon from near mouth of Whitney Creek. U-shaped glaciated form of inner canyon and broad benches on either side, 2000 feet above canyon floor, particularly noteworthy. (U. S. Geological Survey air photo GS-OAL-1-73).

ROCKS

The Sierra Nevada are composed principally of coarse-grained intrusive rocks of several varieties. They constitute the Sierra Nevada batholith, which in truth consists of many separate bodies ranging from 90-150 m.y. old. The rocks in some of these bodies are unusually massive with only a few widely spaced cracks (joints). Such rocks compose the monolithic domes of the Sierras, such as the well-known features in Yosemite.

Since batholithic rocks are intrusive, something older had to be intruded by them. These older rocks are represented by bodies of metamorphics seen along the east side of the range mostly north from Bishop and low on the western slope. They occur largely as dividing partitions (septa) between individual intrusive bodies or as pendants within them. These metamorphics have been formed from great thicknesses of Paleozoic sedimentary and Mesozoic sedimentary and volcanic rocks. Long ago these rocks were tightly compressed into folds trending roughly northwest. Metamorphism occurred both during this deformation and during the subsequent intrusion of igneous bodies. The black and brown colors of these rocks stand out against the lighter shades of the igneous bodies. Any time you see steeply tilted, layered rocks with highly variegated colors in the Sierras, be suspicious that they are metamorphic. Many of the significant metal deposits of the range lie along contacts between igneous and metamorphic rocks.

Cenozoic volcanics constitute the third principal group of Sierran rocks. However, before they were extruded, the Sierra region was strongly uplifted and subjected to extensive erosion which removed an estimated 10 miles of rock, exposing the deep batholithic rocks upon which the volcanics rest.

Volcanic rocks occur in greatest amount, variety, and thickness on the west slope of the range from Sonora Pass north. In the southern Sierra, volcanics occur only in widely scattered spots. On the west slope they appear principally as remnants perched on stream divides in the San Joaquin and Kern river drainages. Some recent flows obstruct the canyon of Golden Trout Creek, and young volcanic cones lie in the middle of the Kern River drainage. Along the east base of the Sierras, volcanic cones, flows, and obsidian domes appear in increasing abundance northward from Little Lake to the Nevada border (Photo 3-21). We will have more to say about these particular volcanics in the Mammoth trip guide.

The youngest deposits of the range, aside from some of the obsidian domes and flows of the Inyo and Mono craters and the associated widespread pumice, are the famous gold-bearing stream gravels of the west slope and the glacial moraines of both the east and west slopes.

STRUCTURE

The Sierra Nevada is usually described as an uplifted fault block tilted west. Although basically correct, this is an oversimplification. There are good reasons for thinking that much of the faulting responsible for the high east-face scarp of the southern Sierras has involved downdropping of Owens Valley more than uplift of the Sierras. The picture currently entertained is this: An area embracing the Sierra Nevada and Owens Valley, and of undefined eastward extent, was warped up into a huge arch attaining its highest elevation about 3-4 m.y. ago. Then Owens Valley started to drop by movement along faults at the east base of the Sierras and the west base of the Inyo Mountains. This occurred during the last 3 million years with a significant amount of the displacement taking place in the last 700,000 years. Indeed, fault scarplets formed as recently as 1872 across alluvial fans, and young volcanics in Owens Valley (Photos 3-28) indicate that the process continues. Faulting and warping has also occurred within the Sierra block itself and, to a minor degree, along the west base, especially near the mouth of Kern River.

You ought to be concerned about what holds the Sierras so majestically above their surroundings. The rocks of the earth's crust are simply not strong enough to support such a massive load. When vibrations generated by earthquakes centered somewhere east of the Sierras arrive at a recording station west of the mountains, it is clear that they have dawdled along the way. They don't travel as fast as they should. Geologists and geophysicists now realize that what slows them down is a gigantic Sierra root, something like a root to a wisdom tooth. The Sierras resemble a great iceberg floating in a sea of rock. The force that buoys them up is the displacement produced by the root, in the same way that the part of a boat below water is what keeps it afloat. Rocks composing the root are similar to the rocks exposed at the surface in the Sierras. They displace heavier, denser materials and transmit earthquake waves more slowly, thus retarding the earthquake waves that pass through the root.

RESOURCES

The Sierras are most famous for gold which occurs in placer gravels on the west slope and in the quartz veins of a metamorphic belt near the west base—the Mother Lode. They have also produced silver, copper, lead, zinc, chromium, and a significant amount of tungsten. Nonmetallic materials include building stones, limestone, and some barite, a barium sulphate mineral used as a source of barium but also valued because of its heavy weight.

SPECIAL FEATURES

The Sierra Nevada are rich in special features, but we select for consideration just one aspect of its scenic development, specifically glaciers. These flowing streams of ice have played a major role in developing the striking scenic character of the High Sierras. Glaciers first formed in these mountains at least 3 million years ago and have been intermittently active until about 10,000 years ago. The small glaciers currently clinging to shaded sides of high peaks are feeble in their erosional, transportational, and depositional effects compared to the giants of the ice ages. Some ice age valley glaciers of the Sierras were 40 miles long and thousands of feet thick.

We tend to think of glaciers too much as scrapers, when in actual fact they are more effective as excavators. Glaciers do wear down rocks by grinding on them, producing in places the eye-catching, scratched, smoothed, and highly polished rock surfaces seen in the higher parts of the range. However, as a factor in erosion this abrasion is much less significant than the capacity that moving ice has for plucking or pulling away (quarrying) big blocks of jointed bedrock. It's a lot easier for ice to pick up a one-ton block of rock and carry it away bodily than it is to laboriously wear it away by sandpapering.

Glacial excavation has created rock basins at the heads of Sierra canyons, now largely occupied by lakes or meadows. Rock basins were also formed farther down canyons where bedrock jointing was favorable. Most glaciated canyons have been widened by plucking, giving them an open U-shape (Photo 2-11) which contrasts with the V-shape of stream-cut canyons. In places deepening and widening of main canyons carrying large glaciers has left tributary canyons "hanging." Today, streams descend from such hanging valleys in a succession of cascades and falls.

Although the floors of many glaciated valleys make pleasant hiking because of their gentle gradients, every now and then one encounters a steep precipitous bedrock rise as much as a thousand feet high. These are glacial steps. They usually occur where a fortuitous juxtaposition of well-jointed and relatively unjointed rock allowed the glacier to exercise great differential plucking.

Excavation at the heads of glaciers undercut the slopes of high peaks and ridges making them ragged, sharp, and narrow. Pointed peaks like the Matterhorn have been given that shape by glacial excavation on at least three sides. Sharpening of divides has not happened everywhere in the Sierras, however, because in some instances ice flowed from the west side of the range eastward through passes across the crest. Such passes are broad, open, and smooth. Perhaps you have hiked through some and have wondered about them.

As the ice streams flowed down canyons they moved into a warmer environment where melting was more rapid, and eventually a balance was struck between the forward movement of the ice and rate of melting. Then the rock debris being carried by the ice was dumped in one place creating an *end moraine*. Melting also occurred along the lateral edges of the ice stream up the valley, and material dumped along the ice margin formed a *lateral moraine*. The Sierra Nevada have beautiful examples of end and lateral moraines, especially along its east base north from Bishop. Here the glaciers pushed out onto the gentle surface at the foot of the range, building huge embankments hundreds of feet high (Photos 3-32, 33). These are located for you in the Mammoth trip guide.

BASIN RANGES

This is rugged desert country with great topographic relief. The lowest point (Death Valley) is 282 feet below sea level, and the highest (White Mountain Peak) is 14,246 feet above. Even the local relief exceeds 11,000 feet, as between Telescope Peak (11,049) and the floor of Death Valley (−282). This means you can keep warm in winter by camping low, and cool in summer by camping high. The Indians knew this long before white man invaded the area.

The name "Basin Ranges" comes about in a curious way. Much of far-western United States is in the Basin and Range province, so named for its long narrow mountain ranges separated by intervening valleys or basins. A subprovince of this region lying between the Sierra Nevada and the Wasatch Mountains, behind Salt Lake City, is designated the Great Basin because it does not drain to the sea. The area of concern here is part of the Great Basin. It is also part of the larger Basin and Range province with long narrow mountain ranges and intervening valleys. Since these mountains lie within the Great Basin, they are designated "Basin Ranges."

CHARACTER AND DIMENSIONS

The triangular area of Basin Ranges in California is bounded on the west by the Sierra Nevada and on the east by the Ne-

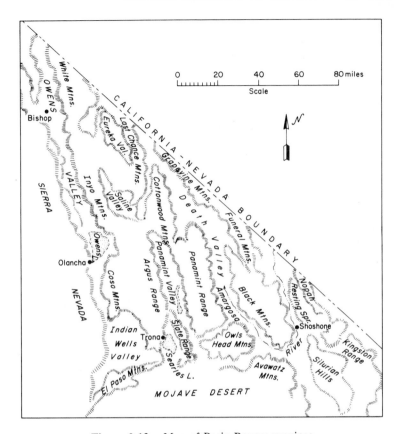

Figure 2-13. Map of Basin Ranges province

vada border. It is cut off abruptly on the south by the Garlock fault (Photo 3-10). This triangle has an east-west base of 135 miles and a north-south height of 170 miles (Figure 2-13).

The mountain ranges and intervening valleys are 50 to 100 miles long by 15 to 20 miles wide, and they trend a few degrees west of north. Roughly half the terrain consists of mountains, which for the most part are dissected and rugged, although some, such as the Panamints, Cosos, and Inyos, have areas of gentle upland. For the most part, the ranges rise abruptly above the floors of intervening valleys with steep, imposing faces. Principal among the ranges are the Inyo, White, Argus, Slate, Panamint, Cottonwood, Last Chance, Grapevine, Funeral, Black, and Amargosa mountains. Major valleys are Owens, Panamint, Saline, Eureka, Death, and Pahrump.

Most valleys are deeply filled with alluvium, and their fans rise gradually and gracefully from central dry-lake flats (playas) to an abrupt contact with the mountains. The ascent of these alluvial slopes gets surprisingly steep toward their heads; so don't assume something has gone wrong with your car if it seems sluggish on the final pitch.

Some valleys have low hills produced by recent deformation or scattered isolated knobs representing older hills only partly buried by alluvium. A few have impressive sand dunes, such as those at the south end of Eureka Valley 700 feet high. Dunes at the north end of Panamint Valley, in Death Valley, and elsewhere are smaller.

The only flowing streams of any size are the Owens and Amargosa rivers, and both end up in local sumps, Owens River (before diversion) in Owens Lake and Amargosa River in Death Valley.

ROCKS

Like Mojave Desert, the Basin Ranges display rocks of many ages and types. There are old-old (early Precambrian) metamorphic rocks, mostly gneiss and schist. Old (late Precambrian) rocks are represented by several thousand feet of weakly metamorphosed sedimentary beds, largely sandstone, shale, conglomerate, and carbonate layers, which are locally intruded by igneous rock to produce the talc deposits widely mined in the Death Valley region. Medium-old (Paleozoic) rocks are abundant and widely distributed. They consist of well-layered sequences of sandstone, shale, and carbonate beds, aggregating tens of thousands of feet in thickness, locally fossiliferous. The not-so-old (Mesozoic) rocks are largely granitic intrusive bodies similar to the igneous rocks of the Sierra Nevada. Small areas of metamorphic rocks of Mesozoic age are known in the south.

Young (Cenozoic) rocks occur as land-laid accumulations of sand, gravel, silt, and fragmental volcanic debris within the valleys and locally in the mountains where they have been uplifted during the process of mountain building. Cenozoic lava flows and associated beds of fragmental volcanic debris are common. The valleys are filled with coarse alluvial deposits, locally uplifted and tilted by very recent deformation.

STRUCTURE

By now you realize that most of the larger topographic features of our western lands are determined by faults and folds. The Basin Ranges are no exception. Each range is a fault block, bounded on one side or the other, often on both, by faults. The ranges have either been uplifted or the intervening valleys dropped down. These movements are geologically young, having occurred mostly in the last few million years, and in places they still go on. This recent movement is indicated by many fresh fault scarplets cutting alluvial fans (Photo 2-12).

Rocks making up the mountain ranges have a complex structure produced by episodes of deformation that occurred before uplift of the ranges. The older rocks buried beneath alluvium in the valleys presumably have similar structural complexities. In the southern part of the Death Valley country are low-angle thrust faults, one of which, the Amargosa, has created a veritable chaos within the rocks involved. (We will have more to say about this Amargosa chaos in the Death Valley trip guide.) Large thrusts also occur elsewhere in the province.

Extending northwesterly for at least 200 miles through the Death Valley region is a major fault structure, the Death Valley-Furnace Creek fault zone (Figure 1-1). It roughly parallels the San Andreas and displays other similarities, such as right lateral displacement and recent activity. Lateral displacement of several miles seems agreed upon, but there is disagreement as to 50 miles. Regardless, this is a major structural feature, and it plays a large role in deter-

Photo 2-12. Fault scarplet up to 50 feet high breaking the surface of Hanaupah Canyon fan on west side of Death Valley. (Photo by John S. Shelton, 4244).

mining major topographic configurations within the region.

RESOURCES

Historically this province has been southern California's major producer of silver, lead, and zinc, primarily because the abundant limestone and dolomite rocks of the area, as intruded by igneous bodies, provide favorable geological conditions for the deposition of minerals containing those metals. Cerro Gordo, high in the southern Inyo Mountains east of Keeler, Santa Rosa Mine, 10 miles south-southeast, and properties in the Darwin district between the Coso

and Argus ranges have been the main sources. In early days the Panamint Range also supplied its share of lead and silver. Tungsten has come from Darwin and the Panamints, and modest amounts of copper, molybdenum, and gold are obtained as byproducts of ore refining for other metals. A little mercury was at one time recovered at Coso Hot Springs.

Among nonmetallic deposits, salines from the dry lakes and from Tertiary sedimentary deposits lead the way. Searles Lake is world famous for the richness and variety of its chemicals, principally salts of potash, boron, and lithium. Some bromine is obtained as well as common salt and the carbonates and sulphates of sodium. Borax was formerly produced in the Death Valley area, and sizeable deposits there may some-day be mined again when richer deposits elsewhere are exhausted. Talc is obtained from mines scattered through the Inyos, Panamints, and the Death Valley area where it has formed in carbonate rocks close to in-trusive igneous bodies. High-grade refrac-tory material used in spark plugs was for-merly mined in the White Mountains, and the Last Chance Range sports a sulphur mine. Mineralization is widespread in this province, and it has been extensively pros-pected, but truly large mining operations have not been developed.

SPECIAL FEATURES

The Basin Ranges are rich in special *features,* such as salt pans, sliding rocks on playas (Racetrack), huge alluvial fans, re-markable turtleback structures (Death Val-ley), volcanic explosion craters (Ubehebe), and sand dunes, among others. Resolutely, we confine attention to an integrated chain of lakes (Figure 2-14) that once occupied parts of some of the principal valleys, 10,000 to at least 100,000 years ago.

During the great ice age the Basin Ranges area was cooler and better watered. The mountains, which even today occasion-ally sport good blankets of snow, were then more heavily covered, and springtime thaws delivered much water to adjoining basins which sustained shallow lakes. However, do not forget our friend, the Sierra Nevada, to the west. This high massive range cap-tured huge quantities of snow during the ice age, and great ice streams tens of miles long choked its valleys. Melting of this ice sent great quantities of water to both east and west. Owens River on the east must have

more than doubled its discharge. It filled the basin of Owens Lake at the south end of Owens Valley to overflowing. At that level, the expanded lake was over 200 feet deep (Photo 3-25) and covered 240 square miles. Its outflow stream ran south through Haiwee Meadows (now Haiwee Reservoir) into Rose Valley and cut the narrow defile at Little Lake forming a narrow gorge with rapids and a spectacular waterfall (*see* field guide, Segment M). Then it emptied into Indian Wells Valley now occupied by Inyo-kern, China Lake, and Ridgecrest. Here a broad shallow lake only 30 feet deep formed before overflow occurred eastward by way of Salt Wells Valley to Searles Basin. The vertical drop from Owens Lake to Searles Basin is over 1500 feet. Waters became deeply ponded in Searles Basin, eventually attaining a depth of 640 feet and backing up Salt Wells Valley into Indian Wells Val-ley to make a single large water body cover-ing some 385 square miles. This was the largest lake in the chain.

At the 640-foot depth, Searles Lake overflowed a divide at the southeast corner into Leach Trough. From here the water flowed eastward down the trough and then made a sharp turn north into the southern end of Panamint Valley. The lake formed in that basin was 60 miles long but only 6-10 miles wide. It covered about 275 square miles and had a depth approaching 1000 feet.

At this level the water overflowed by way of Wingate Pass (Photo 2-13) in the southern Panamint Mountains into Death Valley where it joined the Amargosa and Mojave rivers to make a body over 600 feet deep, named Lake Manly. More attention will be given to Lake Manly in the trip guide to Death Valley.

The shoreline features, cliffs and

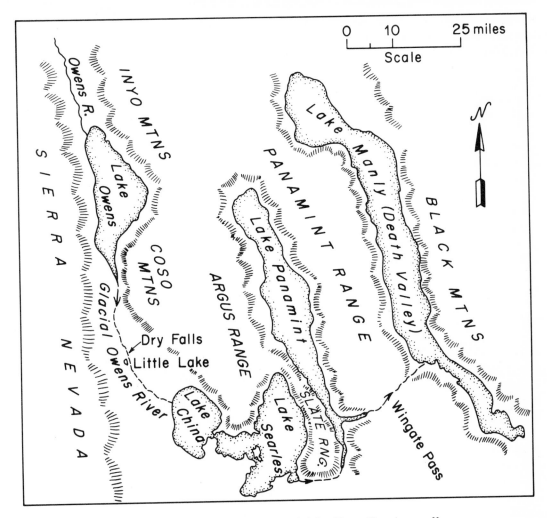

Figure 2-14. Map of pluvial lakes fed by Sierra Nevada runoff.

beaches, of these ancient lakes can be seen today in places around the margins of some of these basins (Photo 3-6). This country must have been extremely attractive when occupied by lakes, and early Indians settled on lake-front sites along their shores.

We still get benefits from these lakes, especially the one in the Searles Basin. They made up a system of huge decanting vessels and evaporation basins. The waters entering Searles Basin had attained a composition such that many chemically valuable salts were precipitated there when the lake dried up. This is the basis of the large

Leach Trough (Garlock F.)
Wingate Pass
San Gabriel Mtns.
Slate Range
Searles Lake
San Bernardino Mtns.
Mormon Pt.
Panamint Mtns.

Copper Cyn.
Panamint Valley

Photo 2-13. Very high-altitude oblique view over central Death Valley, looking south. (U. S. Air Force photo taken for U. S. Geological Survey, 374R-192).

chemical extractive industry operating today at and around Trona. The large calcareous tufa cones and spires, up to 75 feet high, at the south end of Searles Lake, compose one of the weirdest landscapes in all the United States. They are well worth a visit. The road to them is no highway, but it is not dangerous.

CHAPTER

$$3$$

Trip Guides

Reading about geology is all right, but seeing some is more satisfying. This section deals with geological features visible along some southern California highways. Limitations of space prohibit treatment of all the attractive possibilities. Perhaps some future booklet can include more of them. The focus is upon two routes richly endowed with striking geological features and likely to be traveled by people with an innate interest in the out-of-doors. They are San Bernardino to Death Valley, and Greater Los Angeles to Mammoth (Figure 3-1). It was early concluded that the hazards of freeway driving within heavily populated regions make field-trip guides impractical there. The guides therefore start on the periphery of heavily settled areas.

Trips are arranged in segments, each accompanied by a simple sketch map. A good highway map is a desirable supplement, and the county or special area maps issued by automobile clubs are particularly useful. The trip segments can be assembled in various combinations. For instance, a loop trip can be made over the San Gabriel Mountains outward from the Los Angeles area by way of the Angeles Crest and An-geles Forest highways (Segment I) with a return via the Antelope Valley Freeway (Segment J), as shown on Figure 3-1.

On the assumption that most travelers do not care to monitor an odometer (the mileage part of your speedometer), mileage references are kept to a minimum, and points of interest are located wherever possible with respect to prominent roadside features. However, it helps to know that you are 5 miles beyond Mojave or 10 miles north of Victorville; so record odometer readings on the margins of these pages as you pass towns, intersections, or prominent points. Small black triangles on trip-guide margins indicate odometer check points. Odometers cannot be read more accurately than 0.1-0.2 mile. Direction to a geological feature is most easily stated in terms of the hour hand of a clock, with 12 o'clock straight ahead, 9 o'clock to the left, 3 o'clock to the right, and so on. Just imagine a clock face laid down flat in front of you with 12 o'clock directly up the highway. Some locations are numbered thus, ③, on the maps and correspondingly in the trip guides to facilitate correlation. P numbers on maps, as P3-2, indicate location of

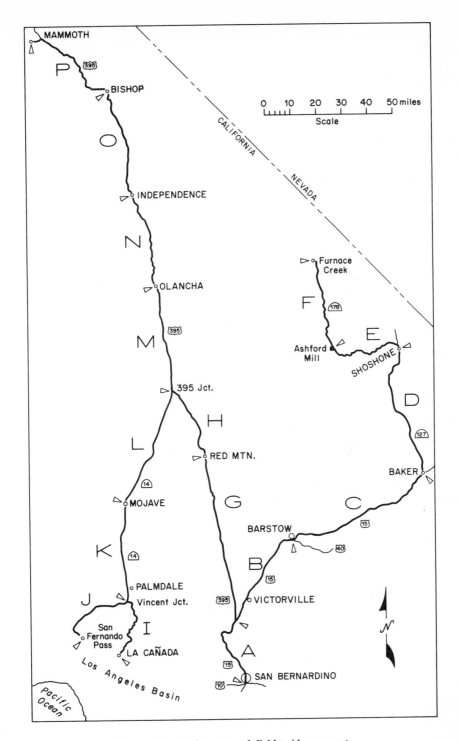

Figure 3-1. Index map of field-guide segments.

roadside photos. These guides are composed for outbound travelers, but with some ingenuity they can be run in reverse. Obviously, they work best for passengers. Visibility and appearances change greatly with time of day, season, and weather. Things clearly seen on the west side of a road in the morning are obscure in afternoon light; so make allowances.

Any highway guide in southern California is likely to be out of date before it can be published. The geology doesn't change that rapidly, but our program of highway construction produces continuing changes in routes and exposures. By the time you use them, some parts of these guides will inevitably pertain to abandoned sections of highways. In some instances, those sections may still be accessible and, being uncrowded, are more fun to travel at a leisurely pace than the parallel high-speed freeways.

Following a field guide requires alertness and skill, and you get better at it with experience. These trips can be repeated several times with profit and enjoyment. Don't be disturbed if you don't see and understand everything the first time. It helps to slow down a little in areas where there is much to be seen, and to make the occasional stops recommended. Try to keep track of check points, know where you are, and occasionally record mileage. In featureless sections of the trip, read ahead in the guide so you can be alerted to items of interest coming up. Study the maps to establish familiarity with the arrangement and distribution of features.

LOS ANGELES BASIN TO DEATH VALLEY

This guide starts at the interchange of Interstate routes 10 and 15 just south of San Bernardino. It proceeds to Furnace Creek Ranch in Death Valley via Barstow, Baker, Shoshone, and State Highway 178 over Salsberry and Jubilee passes and up the valley floor through Badwater, a total distance of roughly 275 miles. It is highly desirable to have read the first two paragraphs of Segment A before getting to the interchange. Things happen rapidly at the start.

Segment A—San Bernardino to U. S. 395 Separation, 31 miles, Figure 3-2

● (Note odometer reading before starting north on Interstate 15.) The San Jacinto fault, historically one of southern California's most active earthquake-generating fractures, passes under the interchange and continues northwestward toward the mouth of Lytle Creek (Figure 3-2). In crossing under the wide sandy bed of Santa Ana River, just northwest of the interchange, the fault forms a barrier which forces underground water to the surface. This is the reason for the perennial flow of water under Interstate 10 just west of the interchange.

● The most recent line of fault displacement lies between the southern pair of three high radio towers seen at 11 o'clock just west of the freeway as one descends from the interchange. It continues past the modern church on the knob at 11 o'clock and extends beyond, out of sight, through the campus of San Bernardino Valley College, where it passes directly under some of the principal buildings.

● In about 4.5 miles our route turns northwestward and ascends the alluvial fan of Cajon Creek for the next 8-9 miles.

● Roughly 7.5 miles from the interchange, we pass through a deep road cut in the northwest end of Shandin Hills where the angular joint blocks of crystalline rocks seen are part of the metamorphic complex.

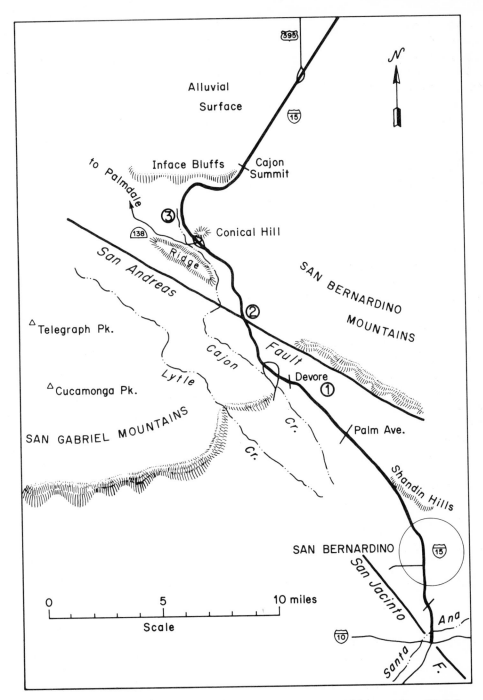

Figure 3-2. Segment A, San Bernardino to U. S. Highway 395 separation. Circled numbers keyed to sections in text.

● Just beyond State College Parkway, and ½ mile away at 1:30 o'clock is a narrow linear ridge, presumably a thin slice between parallel faults. We are obliquely approaching the San Andreas fault which lies along the base of the San Bernardino Mountains to the northeast. Looking to the San Bernardino Mountain front at about 1:30 o'clock, one sees road cuts and other workings related to the outlet of the San Bernardino tunnel and the Devils Canyon power plant of the California Water Project. The tunnel extends south through the mountains from Cedar Springs reservoir at the desert's edge.

● At the Palm Avenue-Kendall Drive exit, a look at 11 o'clock in good weather shows two high skyline peaks in the eastern San Gabriel Mountains, Cucamonga (8859 ft.) in front and Telegraph (8985 ft.) behind. They are composed of igneous and metamorphic rocks.

① Beyond Palm Avenue, alluvial fans extending from the base of the San Bernardino Mountains on the northeast have been trimmed off sharply by Cajon Creek in a 75-100 foot face, ½ mile east of the freeway. This face is well seen a little less than 2 miles from the Palm Avenue exit. Note that the fan surface about half-way to the mountains is here broken by a fault scarp parallel to the mountain front.

► ● (Note your odometer reading under the second overpass at Devore.) Beyond Devore our route parallels Cajon Creek on the southwest and converges slowly with the San Andreas fault on the northeast. The road cuts here would show good sections of fan gravels and highly deformed metamorphic rocks, if they had not been so homogenized by the California Division of Highways. The San Jacinto fault follows a nearly parallel course up Lytle Creek west of Cajon Creek. We are traversing the severely deformed block between it and the San Andreas, two of California's largest and most active faults. About 2.5 miles beyond Devore, a second view of the high peaks is obtained at 10:30 o'clock through a saddle in the Lytle-Cajon creeks divide.

② A little more than 3 miles beyond Devore the highway starts to curve east. We pass through two huge road cuts exposing metamorphic rocks on both sides the road, and suddenly we burst into more open terrain, the Cajon amphitheater. We have intersected the San Andreas fault zone. Look immediately northwest at 9:30-10 o'clock (not the driver) to get a striking view directly up Lone Pine Canyon, a linear cleft defining the trace of the San Andreas fault.

● Within a mile the railroad cuts across Cajon Creek at 9 o'clock exposes light-colored crystalline rocks in contact at the north end with a fault slice of well-layered, dark-brown marine shales of the Martinez Formation (60-70 m.y.). Beyond, in cuts and on hillsides at 9-11 o'clock are extensive pinkish exposures of the Cajon Formation. These upper Miocene (15 m.y.) beds consist of moderately indurated, nonmarine deposits of sandstone, conglomerate, and shale, aggregating a thickness of 9500 feet. They are correlated with similar deposits on the opposite side of the San Andreas fault in the Devils Punch Bowl near Valyermo, 23 miles to the northwest. This is one of several relationships suggesting a lateral displacement on the San Andreas amounting to tens of miles within the last 15 million years.

● At and beyond the Cleghorn Road intersection, look across Cajon Creek at 9:30-10 o'clock to see a high, somber, gray ridge composed of crystalline rocks in back of lower exposures of pinkish Cajon beds. This ridge is a fault wedge, or slice, forming an elongated island of crystalline rock in a sea of Cajon Formation. It parallels the freeway for the next 2 miles.

(3) Passing the Weigh Station, look ahead low down at 11-11:30 o'clock to see large, light colored, flatiron ridges formed by erosion of well cemented, tilted Cajon beds. Beyond the Weigh Station, approaching Palmdale turnoff, the conical hill lying just east of the freeway at 1 o'clock is either a plate of crystalline rock thrust out of the San Andreas fault zone onto the Cajon beds or an island of crystalline rock projecting through Cajon beds which have been deposited around it.

• A mile or two beyond Palmdale exit we start to climb out of the Cajon amphitheater, an open basin created by the erosion of Cajon Creek and its tributaries. It exists because the sedimentary rocks on the northeast side of San Andreas fault are soft and susceptible to rapid erosion, compared to the hard crystalline rocks southwest of the fault. Once Cajon Creek had worked its way across the fault, it had a field day removing the softer materials exposed here.

• Ascending the grade we look north against the high Inface Bluffs carved by the headward growth of Cajon Creek eating back into the desert country. Shortly the freeway swings east and obliquely traverses this face. The upper part of the bluffs consist of fan gravels carried north at earlier periods from the San Gabriel and San Bernardino mountains. These gravels are underlaid by somewhat older terrestrial deposits, including one unit called the Horsethief Formation, which rests upon the Cajon Formation. Unfortunately, the Division of Highways has so homogenized the huge road cuts ahead that we can no longer observe the striking nature of these deposits. We pop over Cajon Summit at 4190 feet at close to 5 miles beyond Palmdale exit.

Our freeway and three railroad lines use Cajon Pass to get out of the Greater Los Angeles area. This route was formed by the erosional work of Cajon Creek taking advantage of the rock and structural patterns provided by the San Andreas fault. Without this unusual erosion the passage from San Bernardino to the desert would be steep and difficult, perhaps impossible for rail lines.

• Once over Cajon Summit we descend a long, gently inclined alluvial slope toward Victorville. This is part of the apron of fans that once extended back to the high mountains on the south before being beheaded and partly eaten away by Cajon Creek.

• About 3 miles beyond Cajon Summit, U. S. 395 separates from Interstate 15. Guide Segments G and H follow it to a junction with State Highway 14 which carries the Los Angeles to Mammouth route description (Figure 3-1). Those going to Death Valley continue on Interstate 15 (Segment B). Those going to Mammoth turn off on U. S. 395 and skip to Segment G.

Segment B—U. S. 395 to Barstow, 46 miles, Figure 3-3

● From the U. S. 395 separation, continue north-northeasterly on Interstate 15 toward Victorville. Mojave River, draining from high, relatively well-watered parts of the San Bernardino Mountains, flows north in a wide sandy bed about 10 miles to the east. We are slowly converging to a rendezvous with it at Victorville.

● Beyond the Hesperia-Phelan exit, the west end of a broad trough extending from Victorville east-southeasterly to the Twenty-Nine Palms country makes the low skyline at 2 o'clock (*see* Mojave Desert province). Ahead are the bedrock hills of Victorville, composed of granitic and metamorphic rocks. Some of the white spots are carbonate-rock quarries (limestone and marble) which provide material for the cement plants of Victorville and Oro Grande. At 10 o'clock are low peaks and ridges of the Shadow Mountains. They contain extensive exposures of metamorphic and igneous rocks like those near Victorville.

● Beyond the Lucerne Valley exit, plumes of dust and smoke from cement plants at Victorville and Oro Grande are often visible. The smoke dead ahead beyond Victorville is from the city dump. At 11 o'clock are granitic rocks of Silver Mountain, and the conical hills at 1 o'clock are erosion residuals carved mostly in metamorphic rocks.

● As the freeway starts to curve around Victorville you will become aware of greater gullying and dissection. This reflects the influence of Mojave River which flows in a course cut about 175 feet below the alluvial surface which we have been traversing.

● Shortly we drop down and cross the river. (Note odometer reading at crossing.) A good stream of water flows here the year around, although the stream bed a few miles upstream is bone dry except during floods. What happens is this: About 1.5 miles upstream at the Apple Valley bridge (Upper Narrows, Figure 3-3) the Mojave River, as it cut down its bed, uncovered a westward projecting ridge of granitic rock buried in the alluvium. The river had no choice but to cut a narrow, steep walled gorge through the granite, creating the Upper Narrows (Photo 3-1). Bedrock close

Photo 3-1. Upper Narrows of Mojave River at Victorville formed as river cut down through a buried granitic rock ridge, now exhumed, as viewed southeastward. San Bernardino Mountains in background. (Photo by John S. Shelton, 3402).

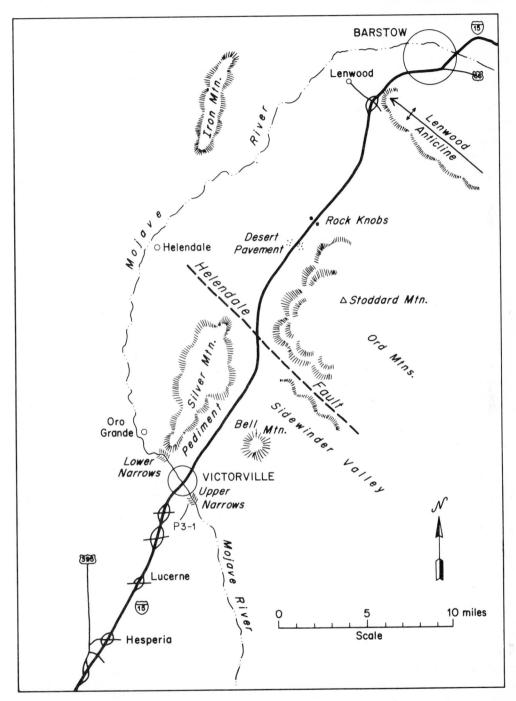

Figure 3-3. Segment B, U. S. Highway 395 separation to Barstow.

under the channel floor at this point brings subsurface water seeping through the porous river sands back to the surface. The water disappears again a mile or two below the Lower Narrow (Figure 3-3), which were formed in the same fashion.

● Across the river we start climbing out of the valley, and good exposures of jointed granitic rock are seen at 1-5 o'clock. They are discolored gray by a mantle of cement dust that has been converted to a hard surface coating. To the left at 9-10 o'clock we see again the granitic rock in Silver Mountain. The gently sloping bedrock surface extending outward (south and southeast) from the mountain base has been formed by erosion and is called a *pediment*. Within 2 miles beyond the river, quarries, roads, and other workings in metamorphosed carbonate rocks (light colored) are seen among hills to the east. At about 4 miles we begin to traverse the broad alluvial surface of Sidewinder Valley.

● Beyond the Speedometer Check sign all but the driver can look back at about 4:30 o'clock for a view of the graceful symmetry of Bell Mountain on the near skyline. The smooth concavity of its slopes is due to the fact that its summit is capped by metavolcanic rocks that yield particles of all sizes upon breakup by weathering. These particles are distributed down the flanks of Bell Mountain in such a manner (coarser near the top and finer near the bottom) as to produce a slope grading smoothly from steep to gentle, hence concave. In another mile or two the rugged Stoddard Mountain mass (4894 ft.) in the Ord Mountains begins to dominate the scene at about 12 o'clock. The rocks therein display considerable variation in color and appearance being largely metavolcanics of possible Triassic age (200 m.y.).

● Shortly the highway curves west; we pass under some large powerlines (note odometer,) and in about 2 miles we start a gentle descent into a shallow valley. We are approaching a crossing of the Helendale fault, one of a dozen major northwest trending faults slicing this part of the Mojave Desert like a loaf of bread (*see* Mojave Desert province). We are here traveling over metavolcanic rocks and debris derived therefrom, and about the only visible indication of a fault is the abrupt change to light colored knobby granitic rock seen a little before crossing the bridge over the shallow wash about 3.5 miles from the powerlines. A few miles to the southeast, and also in Lucerne Valley, Helendale fault is marked by scarplets breaking alluvial fans, an indication of fairly recent movements. (Note odometer reading at bridge.)

● Within the next few miles we get good views of the western Mojave, a region of low relief, broad domes, and small residual rocky peaks and ridges. The Mojave River is in a course about 8 miles west, and we intersect it again at Barstow. The black rocks seen at about 10-10:30 o'clock are part of Iron Mountain. They are dark igneous intrusive rocks relatively rich in iron and magnesium.

● For several miles beyond the bridge, note the blocky, angular character of rock outcroppings and of rock debris on the hill slopes east of the freeway. This is characteristic of the metavolcanic rocks of the Sidewinder Series (Triassic?) largely composing these hills. Locally there are small bodies of intrusive granitic rocks which you may be able to spot by the more rounded nature of their outcrop exposures.

● Between 5 and 6 miles beyond the bridge we come into areas where alluvial surfaces on both sides of the road have patches of desert pavement (smooth, stony, vegetation-free areas). Stones within the pavement are blackened by desert varnish. Desert pavement is a concentration of stones left on the surface of alluvial deposits as finer materials have been carried away. Slight dissection by gullies draining to the intrenched course of Mojave River is slowly destroying this pavement leaving only remnant patches.

● A little short of 8 miles beyond the bridge we pass what looks like piles of huge rounded boulders on both sides of the highway. These are actually the outcrops of jointed granitic bedrock rounded by weathering and granular disintegration, a typical behavior for uniform, coarse-grained igneous rocks in a desert environment.

● Within another mile look dead ahead about 5 miles to a low, gray, skyline ridge with many small gullies. The ridge is a geologically young anticline (an upfold) within relatively unconsolidated alluvial fan materials (fanglomerate). This is the Lenwood anticline, and we will cross its western plunging nose as the freeway curves around into Barstow beyond the Lenwood exit. To complicate matters another northwest trending fault, similar to the Helendale, passes along the southwest margin of the Lenwood anticline.

● Soon the highway bends westward, and buildings at the Lenwood exit are seen ahead. The smooth subdued skyline at 12:30 is the area of Rainbow Basin and the Barstow syncline (a downfold) formed in the famed Barstow Formation, a Miocine (15-20 m.y.), terrestrial basin deposit rich in fossil remains of extinct vertebrate animals (*see* Mojave Desert province, special features).

● Beyond Lenwood exit, the freeway crosses the nose of the Lenwood anticline, and road cuts and hillsides expose some of the gently dipping beds of unconsolidated materials composing this structure.

● Approaching and beyond the first exit to Barstow, at 10-10:30 o'clock, is a prominent reddish-brown rock knob with microwave relay towers and a white letter B. This is a small, cylindrical, intrusive plug of Tertiary igneous rock (rhyolite) formed at a time when the surface volcanics of the Barstow region were being extruded. It has subsequently been etched out by erosion.

Segment C—Barstow to Baker, 61 miles, Figure 3-4

● Just beyond Main Street exit at Barstow, the broad, sandy, and normally dry bed of Mojave River is crossed. During exceptionally wet winters, a good stream of water flows under this bridge. (Note your odometer reading on the bridge.)

● The low hills across the river, traversed for the next 4-5 miles, consist of a mixed and strongly deformed assemblage of Tertiary volcanic and, locally well-bedded, terrestrial sedimentary rocks. At the Bakersfield exit, another large northwest trending fault crosses our route but without recognizable expression at the freeway.

● Descending to the Yermo plain beyond Meridian Road, the Calico Mountains are in view at 9-11 o'clock. The name comes from the variegated appearance produced by complex structural relationships (Photo 1-1) between highly colored sedimentary and volcanic rocks of Tertiary age. The cream and bright-green colors are sedimentary and volcano-sedimentary beds, and the dark reddish-browns are principally volcanics.

● Approaching Ghost Town exit, the large black knob just off the highway at 3 o'clock is Elephant Mountain. This is another near-surface intrusive plug, in this instance of a darker rock than the reddish plug seen in Barstow. (Note odometer at the Ghost Town Road underpass crossing.)

● Beyond Ghost Town Road, look to the mountains at 9:15 o'clock to see the ghost town of Calico, now operated as a tourist attraction by San Bernardino County. Calico was a silver camp from 1882-1896, with a claimed production of 86 million dollars, and then a borax producer until about 1907. The silver deposits were rich, but they pinched out at shallow depths.

The south front of Calico Mountains is bounded by another member of the family of northwest trending faults.

● At the Yermo Exit 1 mile sign, note the highly colored beds in the canyon cut into the mountains at about 9 o'clock.

● About 3.5 miles beyond Ghost Town Road, good bedding is seen in tilted sedimentary and fragmental volcanic deposits of the Calico Mountains face.

● Skip ahead to ① and read about the archeological diggings. Anyone wishing to visit the site should turn off on Mineola Road, about a mile beyond the agriculture inspection station, turn north across the freeway, drive 0.5 mile east on the paved road, go north on the dirt road leading to the County Refuse Disposal Site for 1.1 miles, and then turn east on the one-track labeled road. The public is welcome, and you should find the experience both fascinating and educational.

● Beyond the agricultural inspection station one looks at 2 o'clock down the large trough extending from Barstow to beyond Amboy. This is the route followed by U. S. Highway 66 and the Santa Fe Railway to Needles, and perhaps in earlier times by the Mojave River, which now takes a course closely parallel to the one we travel.

① Just before passing under the large powerlines ahead, look 2 miles north at 9 o'clock. The light area on the hillface about ¾ mile east of the county disposal area (smoke) with trailers and small buildings is near the site of the "Calico Digs," an area excavated by the San Bernardino County Museum of Natural History. At this spot some very primitive "artifacts" have been recovered from deep pits dug in well-cemented fan gravels. If accepted as genuine, they establish occupation of North America by ancient man much earlier than classically pictured, at least 50,000 and perhaps many more years ago.

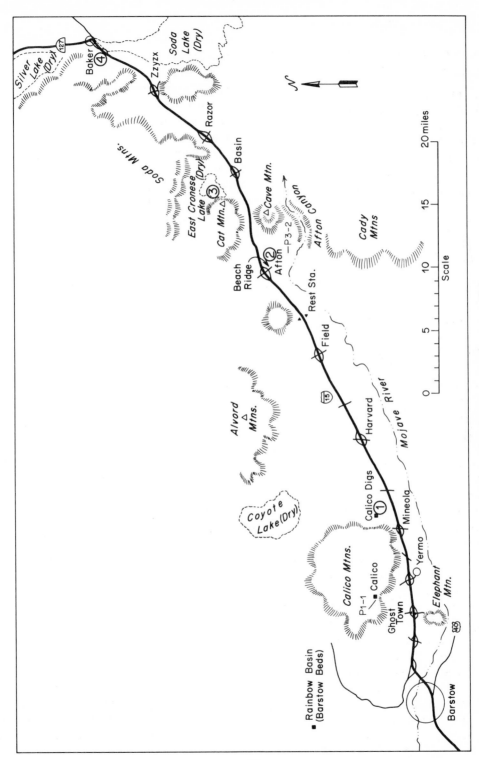

Figure 3-4. Segment C, Barstow to Baker.

● A mile beyond the powerlines, the low hills close to the freeway on the north side expose gently dipping, light-colored, mid-Miocene Barstow beds capped by younger dark alluvial gravels. About 2 miles farther, beyond the next overpass, a view is had of the west end of the Cady Mountains at 1-2 o'clock. The part seen here is composed of dark Miocene volcanic rocks locally mantled by light-colored, windblown sand and silt picked up from the Mojave River alluvial plain by prevailing westerly winds. Little knobs on both sides of the freeway in the next several miles are composed mostly of Barstow beds (light) and volcanic rocks (dark).

▶ ● (Note odometer at the Harvard Road overpass.) About 1.5-2 miles beyond, Alvord Mountains can be seen about 10 miles away at 9-10:30 o'clock. The lighter-colored materials on the skyline near the tower at the east end are young, uplifted, gently deformed fanglomerates. They rest upon much darker Tertiary volcanic rocks of complex structure, which in turn overlie older granitic rocks that have intruded a still older metamorphic sequence exposed mostly toward the west end of the mountains. The irregular light areas within the dark west-end rocks are surficial deposits of wind-blown sand and silt.

● Nearly 5 miles beyond Harvard Road approaching a southward curve, the freeway dips into a broad shallow swale. Here are some dissected soft silty beds deposited in a body of water formerly covering 200-300 square miles in this basin, named Lake Manix. Large bodies of water covered this area at least twice and possibly several more times within the last 15,000-75,000 years when the San Bernardino Mountains had more snow and when the runoff to Mojave River was greater. Wet intervals in desert regions are called pluvial periods, a handy term. Pluvial Lake Manix was a little over 200 feet deep. Its basin was probably created by deformation, perhaps faulting, but it was eventually breached by overflow and downcutting along Afton Canyon.

Animal life was abundant along the shores of Lake Manix. There were shellfish, turtles, beetles, and larger animals including early dogs, bears, cats, mammoths, horses, camels, antelopes, bison, and sheep. Their fossil remains are found in the lake-bed and shore deposits. Among the abundant birds were pelicans and flamingos. The picture of a pink flamingo standing stiffly at attention on one leg in the Mojave Desert is a bit incongruous.

● Approaching the rest area, the prominent mountain at 10 o'clock is a mixture of old granitic and metamorphic rocks.

● Leaving the rest area, the high sharp skyline peak at 1 o'clock is Cave Mountain (3585 ft.). Extensive small-scale gullying of the hills across the valley at 2:30 o'clock suggests relatively unconsolidated deposits, presumably in large part fanglomerates. At a mile beyond the rest area, the commercial plants (1 o'clock) along the railroad (Dunn and New Dunn) are engaged in processing materials trucked out of the surrounding desert.

● In another mile one looks down Afton Canyon at 2 o'clock. As noted, it was cut by the outlet flow from Lake Manix and is now followed by Mojave River and the Union Pacific Railroad. Bedrock in the channel floor forces water to the surface, and short reaches of flowing stream usually exist in the canyon throughout much of the year.

● In 2 more miles, just beyond the Afton Road Exit 1 mile sign, is another and more extensive area of dissected Lake Manix beds. These pale-green, fine silts are capped by younger, brownish gravels.

② Approaching the overpass at Afton Road, the freeway rises and passes through a magnificent gravel beach ridge formed along one of the higher levels of Lake Manix (Photo 3-2). This is a good place to turn off for a look around. The ridge is hard to see from the entrenched freeway route, but after rising to the surface beyond the overpass, look back at 3-4 o'clock to see the backside of the beach ridge and the little playa flat enclosed by it. If you turn off, these relations are easily seen, and the road to Afton Canyon campground at the boulevard stop runs southeast

Photo 3-2. Beach ridge of ancient Lake Manix near Afton Canyon, 2 miles south of Interstate Highway 15, as viewed southward. (From **Geology Illustrated** by John S. Shelton, W. H. Freeman and Company, Copyright © 1966).

along the crest of the beach ridge. (Note odometer reading here.)

• Back on the freeway, starting a mile beyond Afton Road, blocky granitic rock with a modest coating of desert varnish makes up the hill slopes on both sides for the next few miles. Much wind-blown sand and silt from the west have accumulated locally on these slopes.

• In about 5 miles from Afton Road, the freeway starts a descent to Cronese Valley (East Cronese Dry Lake) between Cave Mountain on the south and Cat Mountain on the north. These mountains are composed of an older and tougher type of granitic rock, which is why they are higher, more rugged, and more darkly varnished. Note the very rough, rocky fans along the base of Cave Mountain. The light areas thereon were recently bulldozed during highway construction.

• Nearing the floor of Cronese Valley about 7 miles from Afton Road, note hat rock exposures close to the road on the south are lighter than the rocks higher up the slope. Their desert varnish has been removed by the blasting of wind-blown sand which now partly mantles these lower slopes.

• In crossing the concrete bridge about 8.5 miles from Afton Road, don't be surprised to see a small flowing stream in winter. This is Mojave River water which comes to the surface in Afton Canyon. Upon leaving Afton Canyon the Mojave can flow east to Soda Lake or north to Cronese Lake. At times of flood it often does both. In 1916 flood waters accumulated in East Cronese Lake to a depth of 10 feet.

The abrupt mountain face east of Cronese Valley is the west face of Soda Mountains, an irregular mass extending to Baker. The part viewed here is mostly granitic rock with a mantle of lava on the south flank.

③ Between 1 and 1.5 miles beyond the Basin Road overpass, passengers can look back to 8 o'clock to see a large dune on the east face of Cronese Mountain. It consists of sand blown over the top of the mountain from the west, and the shape viewed in favorable light strongly resembles a cat lying on its tummy with ears and tail visible, hence the names "cat dune" and Cat Mountain.

• Rocks in the mountains around Rasor Road exit are mostly Tertiary volcanics, locally mantled by wind-blown material. About a mile beyond Rasor Road is a good view of Kelso Dunes some 25 miles to the southeast at 3 o'clock. These dunes, fully 500 feet high, lie at the east end of Devil's Playground, a barren sandy windswept plain across which sand is driven for 35 miles from the mouth of Afton Canyon by prevailing westerly winds. The dunes have accumulated at a site where the local topography allows strong storm winds from the north, south, and east to counterbalance the prevailing westerly wind. These beautiful dunes are accessible by road from either Baker or Amboy.

• Looking ahead within the next 2-3 miles, the hills south of the freeway at 12:15-3 o'clock are part of the Soda Mountains, and the rocks are granitic. North of the freeway, at 10:30-12 o'clock, are subdued, lighter-colored hills composed of uplifted young fanglomerates within which are local patches of dark, more highly colored rocks. These dark areas are mostly slices of Jurassic-Triassic (150-200 m.y.) metamorphic rocks inserted into the fanglomerates by faulting within the wide, complex, north-trending Soda-Avawatz fault zone.

• After passing Zzyzx Springs Road, Kelso Dunes are seen again at 2 o'clock in the far distance, and the floor of Soda Lake appears in the mid-foreground. Soda Lake is a remnant of a much larger pluvial water body, Lake Mojave, which lay in this basin 10,000-15,000 years ago.

● In less than a mile the freeway passes through deep cuts in the fanglomerates mentioned above. From the elevated position beyond the cuts one gets good views of Soda Lake and the Devils Playground beyond it at 4 o'clock.

● Coming down the long straightaway toward Baker, the even skyline at 1-1:30 is capped by lava flows, and the dark conical peaks at 2 o'clock are just a few of the twenty-six volcanic cones making up an extensive volcanic field in that area.

④ Take the first turnoff into Baker, and as the highway starts to climb toward the freeway overpass, look to the far (northwest) side of the highway to see a pit in fine bedded gravels of a Lake Mojave beach.

● The dark rock knob just north of the road coming into town is Baker Hill. It is composed of badly broken and deformed upper Paleozoic (Permian) limestone about 250 m.y. old. Note the many small niches and caverns formed by weathering.

● Baker is a good place to gas up and to get supplies before heading north to Death Valley on State Highway 127.

Segment D—Baker to Shoshone, 57 miles, Figure 3-5

▶ • (Note your odometer reading in the center of Baker.) Going north on State Highway 127 toward Shoshone, rocks in the mountain front immediately to the west are early Precambrian metamorphics, probably between 1 and 2 billion years old, and much younger Mesozoic (150 m.y.) granitic intrusives. The light spots mark outcrops of carbonate rock which are not able to preserve a coating of desert varnish. The high Avawatz Range looms on the skyline at 11 o'clock. Rocks in the mountains to the east are also largely early Precambrian metamorphics and local Mesozoic igneous intrusives.

• About 3.5 miles from Baker the floor of Silver Lake playa appears close by on the west and continues for the next 5.5 miles. This is a remnant of ancient Lake Mojave, fed and nourished largely by an expanded discharge from Mojave River. Even now, during very wet winters, Mojave River wa-

ters get this far. The lake flat was flooded in the winter of 1968/69, and in 1916 the basin was filled to a depth of 10 feet by Mojave River floods.

Keep watching the base of the hills along the far (western) shore as you travel north. In places you should be able to make out disconnected parts of a horizontal line (a wave-cut cliff) marking a Lake Mojave water level (Photo 3-3). From 6.5-7 miles out of Baker is a good place to look. Indians favored the shore of this lake, and evidences of their occupation, perhaps 10,000 and more years ago, are found in many sites along this abandoned strand line.

• The piles of light colored material near the road, about 8 miles from Baker, are talc, dumped by trucks hauling from mines to the north. Look west for shoreline remnants as you continue north. Read ahead in ① .

① Approaching the powerline crossing at the north end of Silver Lake, the horizontal mark of the old shoreline is clearly

Photo 3-3. View west across Silver Lake playa. Horizontal line along base of hills marks ancient Lake Mojave water level.

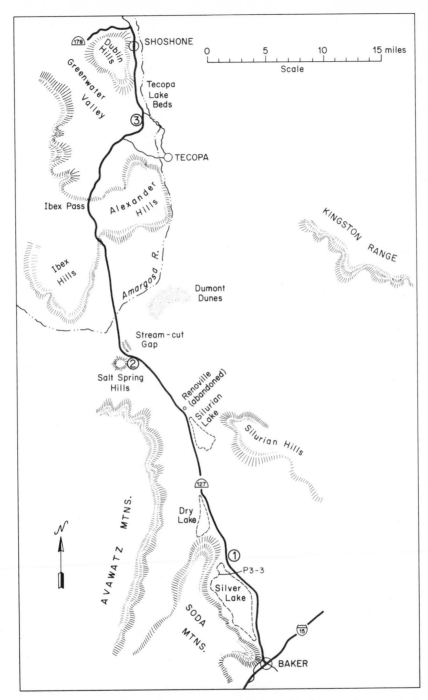

Figure 3-5. Segment D, Baker to Shoshone.

visible in good light along the base of the hills across the lake. The light-colored rock knobs here and at the northwest corner of Silver Lake are composed of a carbonate rock named dolomite, which is like limestone except for a higher magnesium content. Immediately west of the highway, 0.3 mile before passing under the powerlines, is a low beach ridge closing off a small shallow basin on its north side. A good view of relationships is seen by looking back at 7 o'clock from the first powerline crossing. The broad low ridge under the powerline is another slightly older beach. (Note odometer at the powerlines.)

● Within a mile beyond the powerlines, after the first curve west, one sees the Avawatz Mountains at 10 o'clock. Their less rugged southern flank, at 9:30 o'clock, is underlaid by the Avawatz Formation, a lower Pliocene (10 m.y.) fossil-bearing, terrestrial, sedimentary accumulation lying partly within the complex Soda-Avawatz fault zone seen earlier west of Baker. The darker rocks making up most of the range front are early Precambrian metamorphics.

● At 3 to 4 miles beyond the powerline, the Silurian Hills begin to attract attention at 2-2:30 o'clock. They feature an extremely complex structural arrangement of old-old rocks (early Precambrian), old rocks (late Precambrian), medium-old rocks (Paleozoic), not-so-old rocks (Mesozoic), and young rocks (Cenozoic) of igneous, metamorphic, and sedimentary varieties. They will be in view for a good many miles, and occasional glances to the east will give you a sense of their lithologic and structural complexity.

● After taking the second curve, this one to the east, about 4 miles from the powerlines, look at the large alluvial fans built eastward from the base of Avawatz Mountains at 9-11 o'clock. If light is good, several generations of fan surfaces can be distinguished on the basis of differences in degree of darkness, desert pavement development (vegetation-free, smooth areas), and dissection (gullying). Especially good views of these relations are seen at 9:30 o'clock just short of the next curve.

● Beyond this third curve, good views of the Silurian Hills are obtained across the valley at 2 o'clock. On the distant skyline at 1-1:30 o'clock, the high rugged peaks of the Kingston Range loom up. The Kingstons are the home of 7000 feet of weakly metamorphosed, late Precambrian sedimentary rocks, largely shale, sandstone, quartzite, limestone, dolomite, and conglomerate, collectively termed the Pahrump Group, of which we will see something later.

● About 10 miles from the powerlines, Silurian Dry Lake lies immediately east of the highway. Now dead ahead, about 10-12 miles away, are the Dumont dunes (Photo 3-4). They consist of sand piled up by winds blowing from several different directions. The raw alluvium deposited by the Amargosa River is one source of sand. The often snowcapped, distant, skyline mountain behind the north end of the Kingston Range is Charleston Peak (11,918 ft.), located in the Spring Mountains of Nevada.

● Beyond Silurian Lake the road curves west again, and at 11 o'clock, about 5 miles ahead, are the Salt Spring Hills composed of Cambrian (500-600 m.y.) quartzite beds. When closer to these hills note the two different shades of desert varnish, brown and black. The brown variety forms on white to pink quartzite layers; the black occurs on very dark quartzite layers.

② In another 5 miles (17.3 miles from the powerlines), the highway swings west once again, and ahead at 12:15 o'clock is a narrow gap through a rock ridge east of the highway (Photo 3-4). This gap was cut

Dunes
Ibex Hills
Noonday Dolemite
Amargosa River
State Hyw. 127
Dunes
Stream-Cut Gap
Salt Spring Hills
Dumont Dunes
Silurian Dry L.

0   1   2 miles

Photo 3-4. Vertical high-altitude air photo of Salt Spring Hills area, north to top, scale in lower right. (U. S. Air Force photo taken for U. S. Geological Survey, 744V-074).

by overflow from Lake Mojave in pluvial times when Mojave River water ran all the way to Death Valley along this route, joining the Amargosa River ahead. The light-colored, fine-grained deposits and the shore-line features (low bank and horizontal markings) at the base of Salt Spring Hills seen within the next mile indicate that some ponding of waters occurred here before cutting of the gap was completed. At the north end of Salt Spring Hills the highway makes two or three little curves, and when it straightens out, the green trees seen at 1 o'clock mark the location of the gap.

If you would like to have a closer look at this gorge, a rocky but easily passable desert road turns off just short of the little granitic knob west of the highway less than 0.5 mile ahead (19.7 miles from the powerlines). The turnoff is roughly 75 feet beyond the State Highway 127 sign, and the distance to the gap is 0.3 mile. It is interesting to imagine what this spot was like when it harbored a stream powerful enough to cut a gorge into granitic rocks. Even at present, the good growth of salt cedar and cane grass make this spot a bit unusual. East of the gap is a mass of dark, variegated, Cambrian sedimentary rocks, and farther to the east-northeast are exposures of the same Cambrian quartzites seen in Salt Spring Hills.

● Back on State 127 and rounding the corner, we get a good view of granitic rocks at 3 o'clock and of the Cambrian quartzites at 2 o'clock. The southern end of the Death Valley depression lies to the west. A dirt road takes off into it within a mile, at the Harry Wade historical monument, but we will enter from the side some 30 miles north. Small sand dunes are seen east of the road at 1 o'clock from this intersection (Photo 3-4). We come to the Amargosa River (sign) in two miles. Don't be surprised to

see water running here in winter or spring. The river rises in the high Spring Mountain Range in Nevada and at times flows all the way to the salt flats of central Death Valley. (Note odometer reading at the Amargosa ◄ River crossing.)

● After crossing the Amargosa River, the southern end of Ibex Mountains (Saddle Peak Hills) lies at 10-11:45 o'clock. The dark reddish-brown rocks are part of the late Precambrian Pahrump Group of sedimentary beds, and the lighter rock capping the tops of ridges at 11:30 o'clock is the slightly younger Noonday dolomite.

● In another mile the Dumont sand dunes are again in view at 3:15 o'clock.

● At 5.5 miles from Amargosa River, the Noonday dolomite is close by on the west. The subdued hills at 1-3 o'clock are composed largely of fanglomerate. In places this deposit contains lenticular beds composed of large angular fragments of a single type of rock. These probably represent landslide and rockfall accumulations and, being made up of just one kind of rock, are called *monolithologic breccias*. This fanglomerate is similar in character, and possibly equivalent in age, to a late Pliocene-early Pleistocene (2-4 m.y.) deposit of the Death Valley area, the Funeral fanglomerate.

● About 7 miles beyond Amargosa River a microwave relay station west of the highway is passed, and the ascent to Ibex Pass begins. The dark rocks just east of the highway here are volcanics. The deep roadcuts 1.5-2 miles up the grade are in badly fractured granitic rocks, but near the Inyo County line we pass into uplifted fanglomerate deposits containing largely cobbles of granitic and volcanic rocks.

● (Note odometer at the summit just be- ◄ yond the county line.) Descending from Ibex Pass, the northern Ibex Mountains, largely a complex of early and late Precambrian rocks are on the west. The white spots are talc workings. The well-bedded rocks of complex structure at about 10:30 o'clock are Cambrian. Within 1.5 miles from the pass we get good views of the dissected badlands formed in light colored Tecopa lake beds on the valley floor ahead, at 11-1 o'clock. Our old friends the dark Cambrian quartzite beds of Salt Spring Hills are seen in the near ridge at 1 o'clock.

The high ranges beyond the lake basin appear striped because of layering within the thick section of early Paleozoic sedimentary formations composing them. The nearer range is the Resting Spring, and the far skyline range is the Nopah. The combined thickness of beds exposed in these mountains is 23,000 feet, ranging from Cambrian (600 m.y.) to Pennsylvanian (300 m.y.).

● In another mile the Dublin Hills, with beautifully layered Cambrian sedimentary formations, loom up dead ahead. The wide valley at 11 o'clock is Greenwater Valley which we enter farther north. The variegated peak at 10:45 o'clock at the north end of Ibex Mountains is Sheephead Mountain.

● In another 2 miles we are down within the low hills of dissected Tecopa lake beds. A thin layer of younger, dark gravel laid down on top of the soft lake beds, before they were dissected, locally drapes down over the slopes, partly masking the deposits beneath. In places, where the layers of lake silt have considerable coherence, some steep castellated cliffs have developed.

③ About 2 miles beyond the first turnoff (paved) to Tecopa, low, crumbled remains of the adobe walls of the old Amargosa borax works are seen close to the highway on both sides. This site was used during summers from 1882 to 1890 when the heat on the floor of Death Valley prevented crystallization of solutions at the Harmony borax plant.

● Within 0.3 mile beyond the second turnoff to Tecopa (also paved), you can begin to see the faint remains of old, narrow, hand-dug trenches extending up the crests of ridge spurs close to the road on the west. Don't confuse them with the fresher, much wider bulldozer scars. These old, partly infilled trenches, about 2 feet wide and now 1-2 feet deep, were dug during World War I in search of nitrate deposits, the supply from Germany having been cut off. It is amazing that they have survived so long.

● Shoshone is the jumping off place to Death Valley; food, gasoline, water, and a motel are available.

Segment E—Shoshone to Death Valley Floor (Ashford Mill) via Salsberry and Jubilee Passes, 29 miles, Figure 3-6

● The drive from the crest of Salsberry Pass to Furnace Creek Ranch is magnificent; plan to do it at a leisurely pace if you possibly can. In the first mile going north out of Shoshone, the chunks of black rock on slopes immediately west of the highway are derived from a group of lava flows known as the Funeral basalt (1-2 m.y.), and the brownish ridge of well layered rocks 2 miles to the east consists of late Tertiary (3-8 m.y.) volcanic and sedimentary materials. The ridge is part of a fault block lying in front of the darker Resting Spring Range of Cambrian beds (500-600 m.y.). Approaching the Salsberry Pass turnoff (Highway 178) about 1.5 miles out, the abandoned Gertsley borax mine (white spot) is visible near the base of the hills at 2:45 o'clock. (Note odometer reading at turnoff.)

● After turning west on Highway 178, we start to circle the north end of Dublin Hills. In about 1.5 miles the Greenwater Range fills the skyline from 11:30 to nearly 3 o'clock, the part seen here being largely Tertiary volcanics. The far-away sharp peak at 3 o'clock is Eagle Mountain, along the Death Valley Junction road, which is composed of Cambrian sedimentary beds.

● Within another mile we begin to see that the east flank of the Greenwater Range is locally mantled by a thin layer of black rock tilting eastward and resting on top of the more highly colored Tertiary volcanics. These are lava flows of the Funeral basalt, and we see them frequently in the Death Valley country.

● In about 5 miles from the turnoff we are descending gently into the wide Greenwater Valley. Ahead at 11 o'clock is Sheephead Mountain, largely volcanic. In another 2 miles we will be near the center of the valley, a broad synclinal downwarp. We earlier crossed its southern end between the two turnoffs to Tecopa. To the north dark layers of Funeral basalt mantling both its sides are inclined inward toward the valley axis. This is evidence for structural downwarping of geologically recent date, for the Funeral basalt is no more than 1-2 m.y. old.

● In another 2 miles the highly irregular color pattern in the hills at 1:30-2:30 o'clock suggests a complex mixture of rocks. Those exposures are indeed part of a structurally jumbled mass, appropriately named the Amargosa chaos. More specifically they belong to the Calico phase of the chaos, an obviously appropriate name. Other less colorful phases are the Virgin Spring and Jubilee, to be seen ahead. In simplest terms, the chaos phases are breccias formed from shattered and jumbled blocks or sheets of rock shoved out over the ground surface along a series of very gently inclined fractures, called thrust faults. Landslides and rock falls occurring along the freshly created thrust fronts contributed to the complexity of the resulting deposits.

● Volcanic flows and tuffs are seen south of the road ascending to Salsberry Pass. (Note odometer reading at the pass.)

① Within a mile after crossing the pass, we come out onto a fairly smooth alluvial surface. About 2.5 miles from the pass, you might stop to survey the country; geologically, it's a little complex. The dull, dark rocks in the near hills to the south at 9-10 o'clock are early Precambrian metamorphics. The closest rugged dark rock mass at 1 o'clock (Rhodes Hill, Figure 3-6) is early Precambrian gneiss. The more colorful outcrops near its northern base are

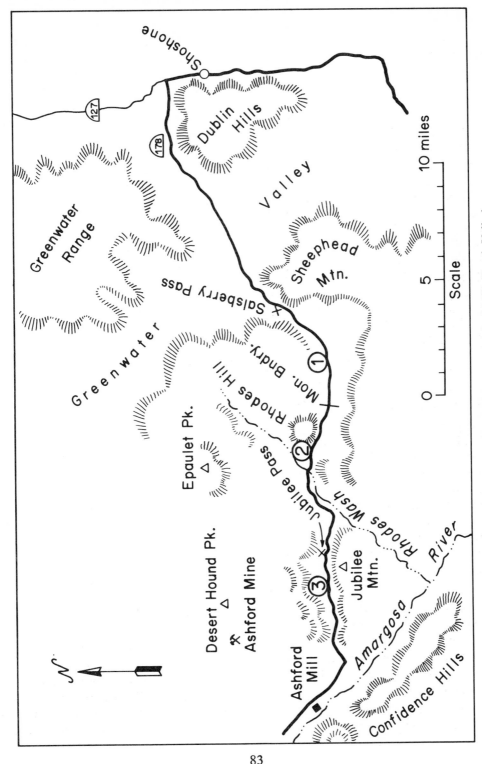

Figure 3-6. Segment E, Shoshone to Ashford Mill (Death Valley).

part of the Virgin Spring phase of the Amargosa chaos. Structurally, the Virgin Spring chaos rests in fault contact upon the gneiss. At 2-2:30 o'clock on the slopes farther back are the highly colored Tertiary rocks of the Calico chaos. The smooth topped skyline hill at 2:15 o'clock (Epaulet Peak) is capped by our friend the Funeral basalt. The highly colored rocks of the sharp skyline peak (Salsberry Peak) at 3:15 o'clock are also part of the Calico chaos.

② About 5.5 miles from Salsberry Pass, beyond Rhodes Hill and past the Monument Boundary, stop short of the little rock knobs just north of the highway. At 2:45 o'clock on the skyline is Epaulet Peak, and on its slopes is Calico chaos. The darker but variegated rocks, still lower at 2:30 o'clock, are Virgin Spring chaos. They rest on some more uniformly dark, greenish-gray, early Precambrian rocks at 2 o'clock which are in the center (core) of an anticline. Low down at 1 o'clock is some more Virgin Spring chaos on the south limb of the anticline. The highly colored rock knob alongside the road ahead is also Virgin Spring chaos. Little knobs and patches of the same chaos are seen resting on early Precambrian rocks just south of the highway along here.

● About 6.4 miles from Salsberry Pass, just south of the highway, is a rocky knob displaying much cavernous weathering. If you stop and walk over, you will find it is composed of broken up (brecciated) carbonate rock (dolomite), probably of Cambrian age. It is part of the Virgin Spring chaos and rests in fault contact with early Precambrian rocks on its west side. In this region, cavernous weathering is a characteristic of chaos rocks.

● In just over 7 miles from Salsberry Pass, where the road first curves south, is a high reddish cliff close on the south side of the road. It exposes chaos resting on weathered, rust-stained, Precambrian rocks. The contact is one of the Amargosa thrust-fault surfaces.

● Swinging around the curve beyond the cliff, the subdued near hills at 2-3 o'clock are Funeral fanglomerate, a late Pliocene-Pleistocene (1-4 m.y.) deposit which is younger than the chaos. The high, pointed, dark peak ahead at 1 o'clock is Jubilee Mountain, composed of a coarse-grained, early Precambrian gneiss.

Looking down the wash ahead (Rhodes Wash), one sees the floor of Death Valley and the rounded whale-back surface of Confidence Hills, a faulted anticlinal structure in soft late Tertiary sedimentary rocks.

● Shortly the highway abandons Rhodes Wash, curves west, and ascends a grade for a mile to the summit of Jubilee Pass. From here the high point on the skyline at 2 o'clock is Desert Hound Peak. The brown to reddish rocks on its lower slopes and to the left are Virgin Spring chaos. The upper part of the peak consists of early Precambrian metamorphics which compose the center (core) of the Desert Hound anticline, a major structure extending north to Mormon Point. (Record odometer reading here.)

③ In about 1.3 miles from Jubilee Pass, close to the road on the north, is a knob of Jubilee phase of the chaos. The large amount of cavernous weathering seen here is characteristic of brecciated rocks. The ridge just beyond, on the south side of the highway, is part of a striking series of tilted red sandstone, fanglomerate, and volcanic layers of Tertiary age. The pink and white smooth areas seen ¼-½ mile south of the highway on this ridge are deposits of volcanic tuff within this sequence.

● Beyond the red sandstone-fanglomerate ridge and just north of the highway, the knobs with good cavernous weathering consist of Jubilee chaos. However, the last big rock knob south of the road, about 3.5 miles from Jubilee Pass, is composed of limestone and dolomite breccia of the Virgin Spring phase of the chaos.

● About 4 miles from Jubilee Pass the southern part of Death Valley is fully in view. Confidence Hills are seen again at 9-12 o'clock, and Shoreline Butte is at 2 o'clock. The horizontal shoreline markings on its slopes become more apparent after we turn north, and they are best seen in late afternoon light. A sharp eyed observer should make out at least a dozen levels.

These strandlines were cut by a lake, nearly 600 feet deep and well over 100 miles long, which lay in Death Valley between 10,000 and 75,000 years ago, named Lake Manly. It was fed by a greater pluvial discharge from the Amargosa-Mojave rivers system and by water that flowed through Wingate Pass (Photo 2-13) into Death Valley from a deep lake in Panamint Valley which was fed largely by runoff from the Sierra Nevada (*see* Basin Ranges, province, special features).

● As we turn north at the road intersection, 4.7 miles from Jubilee Pass, the steep front of the Black Mountains lies to the east. This is a fault scarp composed largely of Virgin Spring chaos as far north as Ashford Canyon (at 1:30 o'clock). Beyond, it consists of early Precambrian rock on the flank of the Desert Hound anticlinal core.

● The ruins of Ashford Mill lie west of the road 2 miles north from this intersection.

Segment F—Ashford Mill to Furnace Creek Ranch, 42 miles, Figure 3-7

▶ • (Note odometer mileage opposite the Ashford Mill turnoff.) Starting 0.5 mile to the north, a succession of Funeral basalt knobs rise above the fan surface just west of the highway. Their north-south alignment suggests that they mark the trace of a fault. To the east, forelying rocky knobs scattered outward from the base of Black Mountains at 2-3 o'clock consist of Jubilee chaos. Horizontal shorelines cut into the northeastern face of Shoreline Butte are usually visible in the first mile or two north from Ashford Mill.

• In a little less than two miles the West Side road (dirt) takes off. This is a good place to stop and look around. At 10:30 o'clock near the center of Death Valley floor is a small cinder cone located on a branch of the Death Valley fault system. On the 9 o'clock skyline is Wingate Pass (Photo 2-13) through which water flowed, perhaps 75,000 years ago, from the large, 1000-foot deep pluvial lake in Panamint Valley. Immediately east of the road to the north, a scarp in black Funeral basalt extends for more than a mile. It marks the trace of another fracture within the Death Valley fault system.

① About 3 miles north from Ashford Mill, with good light, one can see how the little cinder cone, now at 9 o'clock, is sliced apart by right-lateral fault displacement.

• Some 4 miles north of Ashford Mill, at 1-3 o'clock, is a much dissected body of Funeral fanglomerate at the base of the Black Mountains. The contact between fanglomerate and early Precambrian gneiss composing the mountains is determined by the Black Mountains frontal fault. This is a structure consisting of individual fault segments steeply inclined to the west, which are individually linear in trend but which locally diverge from the general north-northwest bearing. Individual fractures or segments of fractures within this zone have recently been active, and we will see many fault scarplets breaking fan surfaces along the mountain base farther north. Some of these scarps are probably no more than a few hundred to a few thousand years old.

• Now is a good time to compare the huge alluvial fans on the west side of Death Valley with the much smaller cones and fans (Photo 3-8) over which we will be driving along the east side. This difference in size reflects in part the greater amount of water and debris discharged by the larger canyons of the higher Panamint Range. However, it is also a product of an eastward tilting of the Panamint-Death Valley block. Geological relationships suggest that eastward tilting has occurred in the immediate past, and tiltmeters measurements on the valley floor show that it continues today. This is the reason the Death Valley salt pan lies so close to the base of the Black Mountains near Badwater. The effect is similar to tilting a saucer partly filled with water.

Tilting has allowed fans on the west side of Death Valley to grow large by extending themselves outward onto the valley floor. At the same time, tilting depresses fans on the east side allowing them to become partly buried by valley-floor deposits, thus reducing their size. The Black Mountains frontal fault marks the eastern edge of the tilting block, and the fault scarplets cutting Black Mountain fans are partly an expression of this movement.

Death Valley is a structural depression. This means that its form has been determined by deformation, probably both warping and faulting. Any closed depression like Death Valley is geologically suspect. Nature

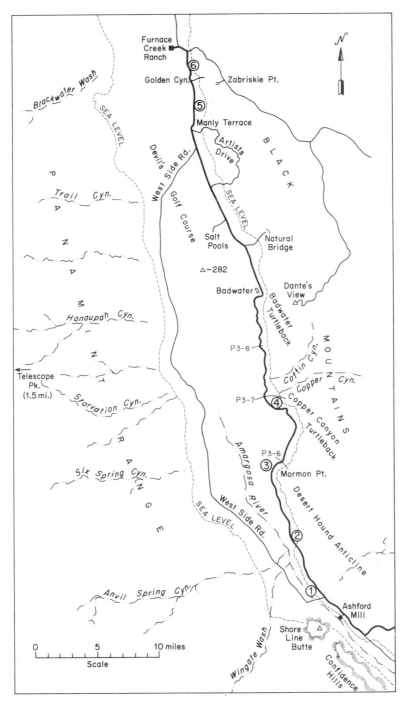

Figure 3-7. Segment F, Ashford Mill to Furnace Creek Ranch.

abhors a vacuum, and she despises closed depressions. She tries to fill them with anything available. In humid areas the initial filling is water, succeeded later by sediment. In dry regions sediment is the principal filling. Even though parts of the Death Valley floor are underlain by as much as 3000 feet of young alluvium resting on top of another 6000 feet of Tertiary sediments and volcanics, nature has not yet been able to complete the filling of this depression, indicating that it is a very young feature and that it has been formed rapidly. With a −282 feet elevation, Death Valley is the drainage sump for a large area in southeastern California and adjacent parts of Nevada. If the climate were more humid, streams would run to Death Valley from all directions, just as they did in pluvial times, and it would harbor a large lake. Death Valley is much better "watered" by springs and streams than most of the surrounding desert because of its topographically low setting.

② The west face of Black Mountains is a youthful fault scarp. Such scarps are locally characterized by wine-glass canyons, and a good example of one is approached at 6 and passed at 7 miles north of Ashford Mill (Photo 3-5). The base of the wine glass is the fan at the foot of the mountain, the stem is the narrow steep-walled gorge cut through the mountain front, and the bowl is the open area of dispersed headwater tributaries.

This reach of the mountain front is made up of Precambrian gneiss and carbonate rocks which yield large fragments. The fan surfaces are therefore rough, irregular, and composed of good sized boulders. At and just north of the wine-glass canyon, watch the toes of some of these fans where they come down close to the salty flats west of the highway. There the boulders develop a decrepit appearance because they are disintegrated by the growth of salt crystals within their pores.

● About 7.5 miles north of Ashford Mill the road curves back toward the mountain front, and directly ahead is a little alluvial cone with a steep, uneven, boulderly surface displaying patchy areas with different shades of desert varnish, from gray to dark brown. This and other cones along the base of Black Mountains have been built largely by a succession of rocky debris flows, and are best termed debris cones. Different flows have inundated parts of this particular cone at widely separated intervals, as indicated by variations in degree of brown varnish development on the surface stones. The grayish lobe in the south central part of the cone marks the most recent flow, possibly less than 100 years old. Note the old, faint, game or Indian trail which crosses the lower part of the fan. It is best preserved in the older and more heavily varnished parts of the cone and has been obliterated by the most recent flows. We don't know much about the rate of desert-varnish formation but suspect it is highly variable, depending upon materials and environment. In some places, perhaps hundreds to a few thousands of years may be required to produce a dark dense varnish under the climatic conditions of the last few thousand years.

● About 9 miles from Ashford Mill, just beyond a spot where the road is crowded against the mountain front by a salt pond, is a little fan with a fault scarp 7-8 feet high breaking its surface. The scarp parallels the mountain front about 20-30 feet from its base. Watch for similar scarplets in fans on up the valley; another one is about 1.5 miles ahead.

Coarse, moderately cemented fan gravels adhere to the base of the mountain front north from here. They have been elevated

Photo 3-5. View north up central Death Valley. (Photo by John S. Shelton, 4238),

(relatively) by movement on the Black Mountains frontal fault and are sharply incised by little stream courses draining from the mountains.

● Looking west at 9:30-10 o'clock across the valley, you can see that the upper parts of fans there are dissected. That dissection is caused by eastward tilting which steepens fan surfaces causing streams to cut down.

③ At 11 miles from Ashford Mill we approach Mormon Point (watch for small sign on left). Mormon Point marks the northern end of the Precambrian core in the Desert Hound anticline earlier seen from Jubilee Pass. At Mormon Point the range front is set back to the east owing to complexities in the frontal fault system. (Stop at the Mormon Point sign and record odometer reading.) At about 12 o'clock a look northward along the face of the mountains reveals a change in appearance and character where Precambrian rocks give way to less resistant, less homogeneous, and more highly colored Tertiary volcanic and sedimentary rocks.

● As the road curves back toward the mountains, 0.5 mile from Mormon Point, a mass of Funeral fanglomerate straight ahead lies in fault contact with early Precambrian schist, gneiss, and marble which compose the high mountain face behind. Younger fan and lake-shore gravels locally mantle the north slope of the Mormon Point peninsula (Photo 3-6). Keep your eyes peeled

Photo 3-6. Horizontal lake shorelines cut in gently inclined Pleistocene fanglomerate deposits behind Mormon Point, view to south. (Photo by John S. Shelton, 3459).

for small fault scarplets cutting across alluvial fans at the head of the Mormon Point re-entrant.

● In about 1.5 miles from Mormon Point, the road straightens out on a north-northeast course. Here we look dead ahead to the southwest limb of the Copper Canyon turtleback. The nose of the structure is at the Precambrian-Tertiary contact ahead at about 11:30 o'clock.

A turtleback is an unusual geological structure. It consists basically of a mass of Precambrian rock in the core of a plunging anticline which has been exposed by erosional removal of overlying deposits. As a topographic feature it has some resemblance to the shell of a turtle. Where remnants of the overlying deposits remain, they are seen to be in fault contact with the Precambrian core of the fold. According to some interpretations, faulting occurred before the folding; according to others, it is considered essentially a contemporaneous event. The turtleback is composed of the Precambrian rocks underlying the anticlinally folded fault surface. These structural relationships are best seen a little south of the mouth of Copper Canyon ahead, and we view another turtleback north of Badwater. Mormon Point is also a turtleback, but its geological relationships are not as clearly seen.

● The bouldery fans seen along the straightaway here are composed of fragments of early Precambrian gneiss.

● About 5 miles from Mormon Point, just where the road starts to curve west around the large Copper Canyon fan, several little debris cones against the Black Mountains base display prominent patches with different degrees of desert varnish (Photo 3-7.)

④ Stop at something less than 0.5 mile out onto the gently sloping Copper Canyon fan and look around. In the mountain front to the east are gray Precambrian metamorphic rocks in the core of the Copper Canyon turtleback. They are overlain by brownish and red beds of the Pliocene (10 m.y.) Copper Canyon conglomerates (Photo 3-7). If the light is right as you continue north, you will be able to see layering in these conglomerates dipping directly into the Precambrian rocks, indicating a structural discontinuity (fault) between these units. At 3:30-4:30 o'clock at the mountain base are the varnished debris cones, one of which has patches of four different degrees of varnish on its surface. The horizontal lines on the mountain face, a little above the cones, especially at 4:30 o'clock, mark old lake shorelines. The narrow slot at the mouth of Copper Canyon, at 2 o'clock, is partly obscured from this view by a fault scarp about 75 feet high in gray fan gravels. A look back toward Mormon Point should show, in reasonable light, lake shorelines cut into fanglomerates (Photo 3-6).

● About 8.7 miles from Mormon Point, the highway heads westward again as it swings out to round the next fan (Coffin Canyon). On the skyline at 11:30 o'clock is Telescope Peak (11,049), often snow-capped in winter. It is composed of late Precambrian, weakly metamorphosed sedimentary rocks.

From here to Badwater the road swings out and in over five alluvial fans, some so youthful and perfect in form as to look almost artificial (Photo 3-8). Their size varies with the area of drainage within the mountains and the ease with which the bedrock therein is eroded. Their steepness strongly reflects the coarseness of debris (stone size) composing them. Copper and Coffin canyon fans have gentle slopes because the Tertiary rocks in their drainage basins yield much fine material. Closer to Badwater we

Mouth of Copper Cyn.

Up Faulted Fan Surface

Copper Cyn. Tertiary Beds

Fault Contact

Varnished Debris Cone

Turtleback Core (Precambrian)

Photo 3-7. Copper Canyon turtleback, viewed from west. (Photo by John S. Shelton, 4241).

Photo 3-8. Small alluvial fans at base of Black Mountains north of Coffin Canyon (Photo by John S. Shelton, 3441).

cross rougher, steeper, more bouldery fans derived from tough Precambrian metamorphic rocks.

● Between fans where the highway closely approaches the mountain front, for example at 10.1 miles from Mormon Point and also a mile farther north, you can see that the rocks within the frontal fault zone are badly chewed up. Frankly, they are a mess.

● If traveling in the late afternoon, look across the valley to the base of fans on the west side. You may be able to see the shadowed face of a long, low fault scarp at about 11 o'clock cutting across the lower part of Hanaupah Canyon fan (Photo 2-12). Don't mistake lines of vegetation for the scarp.

● At 12.2 miles from Mormon Point, a segment of steep mountain front has broken loose in the form of a rock slide. You may

be able to recognize it from the jumble of huge blocks on the lower part of the mountain face.

● In another mile, remnants of fan surfaces at two levels above a fault scarp, at the mouth of a canyon, indicate two successive uplifts. A mile beyond in a similar setting, gravels on a single uplifted surface are heavily covered with desert varnish, which indicates that they have long been undisturbed compared to those on the present fan surface.

● By the time you have reached 15 miles from Mormon Point, you have probably become aware that the salt flats has begun to hug the base of the Black Mountains. This occurs because the eastward tilting of Death Valley is particularly marked here. The character of the pan surface changes with the supply of water, but in places it is often broken into irregular, polygonal fragments 5-15 feet in diameter with turned up edges. They look something like the ice floes in the classical painting of Washington crossing the Delaware. These polygonal plates seem to grow at their edges by crystallization of salt in cracks. The breakup of boulders at the toes of fans by salt-crystal growth is again well seen just east of the road in this reach.

● Badwater is a good place to stop. North along the range front at about sea level height (*see* sign on the mountain face) are little remnants of cemented gravel adhering to the mountain front. At least some of them mark the shoreline levels of a pluvial lake. Since Lake Manly was a good 600 feet deep at maximum, these deposits must represent a lower and later stage when the water depth was only about 300 feet. Small remnants of well-worn shoreline gravels lie at higher levels, but they cannot be identified from the road.

If you walk out on the salt pan, you will see the shorelines and other relationships more clearly. Structurally, we are viewing the west flank of the Badwater turtleback sliced off longitudinally by the Black Mountains frontal fault. The nose of the turtleback is seen better a little farther north. The surfaces of fans near Badwater are broken by fault scarplets, especially the fan to the south. (Before leaving Badwater note odometer reading.)

● Leaving Badwater we move north along the west flank of the Precambrian core of Badwater turtleback. Within the first mile, considerable salt disintegration of stones has occurred just east of the road. Remnants of gravels cemented by calcareous deposits are seen at about sea level height on the mountain face.

● In little over 2 miles the nose of the turtleback becomes more apparent at about 2 o'clock. The light-colored rocks are the Tertiary deposits lying above the anticlinally folded turtleback-fault surface. Looking to about 3:15 o'clock you may be able to see a little conical knob of Tertiary rocks still resting on the crest of the turtleback. It is a remnant of the former covering of Tertiary deposits.

● In about 3.5 miles is the turnoff to Natural Bridge, a 50-foot span in sedimentary rocks of the Tertiary Artists Drive Formation. A good view of Badwater turtleback is seen from here. Northward the mountains are composed of Tertiary volcanic and sedimentary rocks aggregating a thickness of 13,000 feet. The Black Mountains contain a surprisingly small amount of *in situ* Paleozoic rock, in view of tens of thousands of feet of Paleozoic sedimentary beds exposed in adjacent areas. It appears that the Black Mountains block, lying between two large fault systems, the Death Valley on the west and the Furnace Creek on the east, has been strongly uplifted and deeply eroded sometime in the past. The total cumulative uplift and erosion must have amounted to at least 25,000 feet in order to remove the thick Paleozoic section which almost surely once covered this area.

● North of Natural Bridge turnoff is the beginnings of a group of low hills lying in front of the main Black Mountains mass. They are made up largely of highly colored Tertiary volcanic rocks with lesser amounts of associated sedimentary deposits, all rather highly deformed. These hills are part of a subsidiary fault block lying between the Black Mountains block and the Death Valley block (Photo 3-9).

● About 4.5 miles north of Badwater our highway starts across an alluvial fan of no-

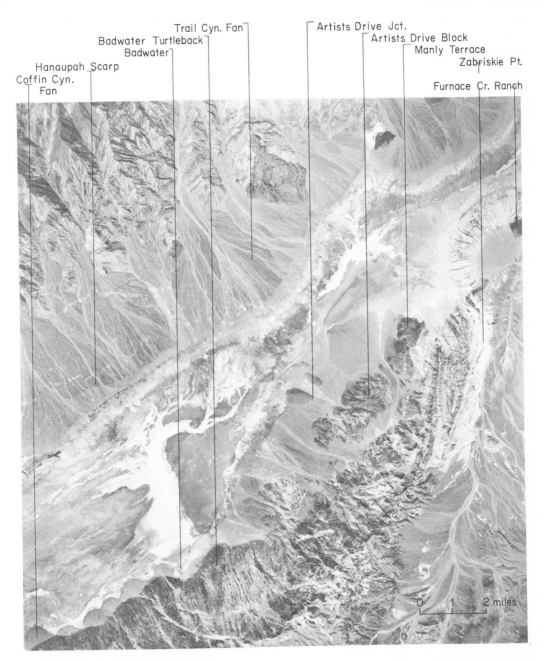

Photo 3-9. High-altitude vertical photo of central **Death Valley**, north toward upper right corner, scale at lower right. (U. S. Air Force photo taken for U. S. Geological Survey, 374V-192).

tably gentle slope. Although large stones locally dot its surface, the fan contains much fine material derived from the Tertiary rocks composing the Artists Drive block. This is the principal reason for its gentle slope and smoothness.

• In less than a mile is the Devils Golf Course (salt pools) turnoff. A visit there will give you a look at the extremely rough surface of a relatively pure salt pan. This relief is a product of solution and of localized crystallization along fractures forming ridges and spires of solid salt.

• In another 2.5 miles is Artists Drive turnoff. Artists Drive is a one-way, 9.5-mile road through a highly dissected badland with narrow washes cut into highly colored sedimentary and volcanic rocks of the Funeral fanglomerate and Artists Drive Formation.

Just west of this turnoff are two low ridges of Funeral fanglomerate warped up by stresses developed through lateral displacement along fractures of the Death Valley fault system. If you feel a need to stretch your legs, get out and walk to the top of the ridge opposite the turnoff. Here you will find volcanic boulders deeply scoured by the blasting of wind-blown sand.

• Within the next mile look ahead at 1 o'clock to black hills of Funeral basalt at the base of the mountains. A close inspection should reveal faint horizontal scars of lake shorelines on their slopes. The nearly flat top of this projecting point is also a lake formed feature called Manly Terrace (Photo 3-9). It may have been the site of very early Indian occupation of Death Valley.

• About 10 miles north of Badwater and a mile short of the junction with West Side road, excellent stratification can be seen in the rocks composing the Black Mountain front. The great variety of colors displayed (tan, cream, lavender, green, brown, gray,

and white) suggest that these deposits contain much fragmental volcanic debris. West Side road comes in about 10.6 miles from Badwater.

• (Note your odometer reading at Artists Drive exit point.) The black rocks at the base of the Panamint Range (10-11 o'clock) are Funeral basalt.

• We round the point of Manly Terrace, pass Mushroom Rock, and about 1.5-2 miles from Artists Drive exit we get a good view of Furnace Creek beds which here make up the Black Mountains front. The Furnace Creek is a Pliocene (3-12 m.y.) formation, consisting mostly of light-colored, soft, silty lake beds with intercalated conglomerate and volcanic layers, 5000 feet in aggregate thickness. They are beautifully seen from Zabriskie Point off the Furnace Creek Wash road. These beds contain considerable borax, and these hills are dotted with old abandoned borax mines and prospects.

⑤ About 2.1 miles from Artists Drive exit we come to a Dip sign and a 100-foot section of concrete pavement. The fan surface here is scoured, dotted with large boulders, and locally spotted with polygonally cracked deposits of dry mud, all suggestive of flooding. West of the highway, a vertical-walled gulch cut into the fan deepens to more than 20 feet near the mountains. This fan is fed by Gower Gulch which heads at the Zabriskie Point overlook. The flooding and dissection are the result of artificial diversion of Furnace Creek drainage into the head of Gower Gulch. This was done to protect buildings in lower Furnace Creek Wash and on its fan. However, it would have been only a matter of time until the diversion would have occurred naturally, for headward working Gower Gulch was on the verge of capturing Furnace Creek. Stream capture is a common geological phenomenon. It produces anomalous drainage patterns and effects changes in the regime of both the captured and capturing stream.

● Roughly 3 miles from Artists Drive exit is Golden Canyon, an interesting little drive up a steep walled gulch cut in Furnace Creek beds.

⑥ At 3.7 miles from Artists Drive exit, and extending for 0.4 mile, we begin to pass a beautiful little fault scarp, 2-7 feet high, breaking the fan surface 20-100 feet east of the road. Some of the gravels in the face of the scarp look different from those on the fan surfaces because the fault displacement has brought up gray gravels of the larger Furnace Creek fan which here underlie a thin deposit of brownish debris derived from the hills immediately to the east.

● Approaching Furnace Creek Inn, the steeply tilted, greenish beds seen are fanglomerates in the Furnace Creek Formation. We turn left on Highway 190 and proceed down the surface of Furnace Creek fan to Furnace Creek Ranch.

LOS ANGELES BASIN TO MAMMOTH

These guide segments are arranged so that travelers bound for Mammoth can depart the southern California metropolitan areas by three routes: San Bernardino, via Interstate 15 and U. S. 395 (Segments A, G, and H), La Cañada, via Angeles Crest and Angeles Forest Highways (Segment I), and San Fernando Pass, via State Highway 14 and Antelope Valley Freeway (Segment J). The San Fernando Pass and La Cañada routes join at Vincent Junction near Palmdale, and they subsequently rendezvous with the San Bernardino-U. S. 395 route in Indian Wells Valley. Travelers using this last exit follow the initial leg (Segment A) of the Death Valley trip as far as the U. S. 395 separation. From Indian Wells Valley the consolidated route proceeds up Owens Valley and through Bishop to Mammoth.

San Bernardino Exit

Use the introductory section and Segment A of the Death Valley trip as far as the U. S. Highway 395 separation, then proceed with Segments G and H.

Segment G—Junction Interstate 15 to Red Mountain via U. S. 395, 70 miles, Figure 3-8

● This segment crosses the eastern part of the western Mojave Desert where topography is subdued and rock exposures are few. This provides opportunity to relax and familiarize yourself with areas ahead where features of interest occasionally come in rapid succession. (Record your odometer reading upon turning off onto U. S. 395.)

● In 2 miles we descend into and climb out of a deep, flat-floored gully (Oro Grande Wash). This is but one of a family of similar gullies cut into the alluvial apron sloping north from Cajon Summit. The cutting had to occur while streams from the mountains could still flow north, and that was before excavation of the amphitheater by Cajon Creek. It suggests uplift of the region while the alluvial apron was still intact. Repeated surveys of bench marks along railroad lines through Cajon Pass indicate that the area is still rising at rates between 0.5 and 2.5 feet per century.

● In another 1.5 miles we cross the east branch of the California Aqueduct which carries water to the Cedar Springs reservoir at the north base of San Bernardino Mountains.

● Going north from the aqueduct, if visibility is at all passable, we get a good view of the Victorville area at 1:30 o'clock with its granitic- and metamorphic-rock hills. The light spots are mostly carbonate-rock quarries supplying material to the cement plants of Oro Grande and Victorville. In 1970, cement was the third most valuable natural resource produced in California, after oil and natural gas.

Looking back to 7-8 o'clock gives a good view of the San Gabriel Mountains, especially striking when snowcapped in winter. The light spot at 8 o'clock is a rock quarry. Eastward at 3 o'clock, the low saddle on the eastern skyline marks the Victorville-Twenty Nine Palms trough (*see* Mojave Desert province), and at 4 o'clock are the San Bernardino Mountains.

● Starting at about 17 miles from turnoff and extending for 10 miles the hills and ridges on the skyline at 9-11 o'clock are the Shadow Mountains. They are composed of granitic rocks intruding a metamorphic sequence of complex structure and a wide variety of rock types, including marble, schist, quartzite, and a varied series of metavolcanics. The light colored isolated

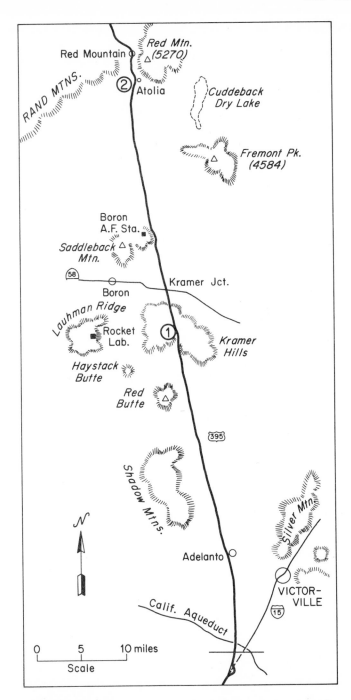

Figure 3-8. Segment G, U. S. 395 separation to Red Mountain.

hill at the north end, seen at 9:15 o'clock at the Shadow Mountains Road crossing 24 miles from turnoff, is Silver Peak (4043 ft.).

● Some 29-30 miles from the turnoff, at 10 o'clock, are Red Buttes, composed of volcanic rocks. Ahead are the Kramer Hills, and within another 4 miles we start through them.

① That part of Kramer Hills seen close along the highway is composed largely of mid-Tertiary (15-30 m.y.) sedimentary and volcanic rocks of the Tropico Group, named from Tropico Mine about 40 miles to the west. This group rests upon older granitic rocks and is locally capped by fanglomerates. The first ridge, just west of the highway, exposes white, well-layered, silica-rich lake beds overlaid by darker lavas, both gently dipping. This sequence is representative of the Tropico Group in this area, and is seen again in the mid-distance east of the highway in another mile. In less than 2 miles we start passing through some large road cuts. The first one exposes coarse fanglomerates, the next one dark lavas, and the third one shows a different type of bouldery fanglomerate.

● Shortly we come to Kramer Junction. (Note your odometer here.) About 6 miles west is the settlement of Boron with its associated borate-mineral mines (*see* resources section, Mojave Desert province). Borate minerals were discovered in the area during drilling of a water well in 1913. Further exploratory drilling in 1925 by Pacific Coast Borax Company revealed larger and more desirable borate deposits at depths of less than 400 feet, and in 1926 sinking of a shaft led to development of the area into what is now the world's most productive borate deposit. In 1957 a huge open pit was developed in the shallower deposits at the west end of the district. The key to the operation here is not only the shallowness and richness of the deposits, but also the fact that much of the boron is in sodium borate compounds, rather than the more widely distributed calcium borates. Sodium borates are more easily and cheaply refined and marketed. The borate minerals occur as layers, lenses, nodules, and veinlets within lake-clay beds (shales) within the upper Tropico Group (15-20 m.y.). The open pit a few miles northwest of Boron is worth seeing. There is a viewing point, and visitors are welcome.

● Going north out of Kramer Junction, the high hill about 4 miles off the road at 10:30 o'clock is Saddleback Mountain, obviously named from its shape. It is largely granite with a capping of younger lavas which shed a mantle of black debris over the slopes.

● In 6 miles we pass Boron Air Force station, a tracking operation situated on top of a low granitic ridge.

● For the next 15 miles we traverse a monotonous stretch of plains and low rounded knobs. The surface material is alluvium and disintegrated granite, but granitic bedrock lies not far beneath the surface over much of this country. Some granites disintegrate so readily in a desert environment that they underlie relatively featureless terrain.

● At about 10 miles from Kramer Junction on the high skyline at 2:30 o'clock is Fremont Peak (4584 ft.) composed of old (Precambrian) granitic rocks intruded by much younger (Mesozoic) granitic rocks.

● About 15 miles out of Kramer Junction the Rand Mountains (granite and schist) show up well on the skyline from 10-12 o'clock, and Red Mountain (5261 ft.), a volcanic knob, is prominent at 1 o'clock.

② About 21 miles from Kramer Junction the road makes a major curve east, and we are on the approach to Atolia. Beyond the curve in the near distance at 10:30 o'clock are piles of gravel representing old placer workings for tungsten. We will see similar piles just east of the highway coming into Atolia, 2 miles ahead. These workings are in alluvial gravels where pieces of a heavy, tough, resistant tungsten mineral, scheelite ($CaWO_4$), have accumulated. The Atolia district has been a major tungsten producer.

● As we pass through the town of Atolia, at about 23 miles from Kramer Junction, look east at 3:15 o'clock to see Cuddeback Dry Lake about 7 miles away.

● Within 3 miles we pass through the town of Red Mountain. Red Mountain and Atolia, along with Johannesburg and Randsburg just ahead, compose one of California's more famous mining districts. The area was started as a gold camp in 1895, and gold was the chief product up to World War I. The war stimulated production of tungsten in the Atolia area, and after the war artificially high prices excited interest in silver, resulting in discovery of rich bonanza deposits near Red Mountain. The total production of metals between 1895 and 1924 exceeded 35 million dollars worth, and total production from the district may approach 50 million dollars, about equally divided among gold, silver, and tungsten. Gold and tungsten production continued at intervals up to the mid-1950's, usually in response to temporarily high prices, especially for tungsten, during World War II and the Korean conflict.

The ore deposits were shallow, the deepest workings being 600 feet, and some were fabulously rich. The gold is associated mostly with quartz veins in granitic rocks and the Rand Schist, a Precambrian (?) formation that we will see shortly. The tungsten occurs as veins in younger (Mesozoic) granitic rock around Atolia and Red Mountain and as placer deposits in alluvial gravels. The silver is in veins in the Rand Schist near Red Mountain. Although occurring in old rocks, the ore deposits are relatively young, having formed during an episode of Miocene (12-25 m.y.) surface and near-surface igneous (volcanic) activity.

*Segment H—Red Mountain to Junction with
State Highway 14, 29 miles, Figure 3-9*

▶ • (Record odometer in Red Mountain.)
Within 2 miles we pass through Johannesburg; Randsburg is a short distance west.

• Between Red Mountain and Johannesburg the piles of coarse, multi-colored rock debris on the hillsides are waste dumps from mines. The piles of fine material are tailings (ground up rocks) from mills that processed the ore.

• A mile beyond Johannesburg we pass the Randsburg turnoff, and soon the highway straightens out on a north-northwesterly course downhill. El Paso Range lies across the valley at 9-11 o'clock, and Summit Range is at 11-12:30 o'clock. If weather is clear, you may just see a little of the southern Sierra Nevada crest on the far skyline at 11 o'clock. In about a mile the view westward at 9 o'clock is along Cantil Valley to Koehn Dry Lake.

• The slopes on either side of the road are partly mantled with slabs of Rand schist; you occasionally catch flashes of reflected light from the smooth fragments. White areas are the outcroppings of quartz veins and fragments derived therefrom. The first road cut, a little more than 1.5 miles from the Randsburg turnoff, exposes a cross section through the schist.

• At the bottom we cross railroad tracks, the highway curves a bit east, and in less than a mile the Garlock road (paved) joins from the west. Upgrade from this intersection, the dark rocks capping ridges on both sides and shedding boulders downslope are lavas.

① As we go uphill a 50-foot scarp, dappled in gray and white, approaches the highway obliquely from the west. The scarp marks the trace of the Garlock fault (Photo 3-10), a major left-lateral fracture

(*see* Mojave Desert province). Near the top of the ascent (8.5 miles from Red Mountain) where the highway curves east and then west, and an abandoned curve in the highway takes off east, we intersect the fault. Fanglomerates exposed in the road cuts on both sides are badly chewed up, and within the last two or three years cracks have developed here in the newly surfaced highway, suggesting slow drift (creep) along the Garlock fault in this area.

• The road to Trona takes off east in 1.5 miles. A distant view of Telescope Peak in the Panamint Range, often snowcapped in winter, is seen at 2 o'clock.

② In another 3.5 miles we again cross the railroad and pass into a terrain made up largely of old messy granitic rocks that display different colorations reflecting secondary alteration by mineral solutions. Such alterations are often associated with ore deposits, and that is one reason why this area is "gopher holed" with prospects.

• In another 3.5 miles from the last railroad crossing, we pass Ridgecrest Road turnoff (Photo 3-10) and start down a long gentle straightaway headed northwest. (Note odometer here.) We see Indian Wells Valley, and the southern Sierra Nevada fill the skyline from 9-12 o'clock; Owens Peak is at 11 o'clock, Olancha Peak is at 12:30 o'clock, and the Argus Range (largely granitic rocks) composes the skyline at 2-3 o'clock. Within 2 or 3 miles, look at 9 o'clock to see lava flows capping Black Mountain.

• In about 5-6 miles we see the southern Sierras more plainly and can recognize that the light colored, clean, hard granitic rocks of their core have a different topographic expression than the more easily weathered granitic and metamorphic rocks composing the range front. The line of the new (1970) Los Angeles Aqueduct across the lower face of the Sierras is plainly visible.

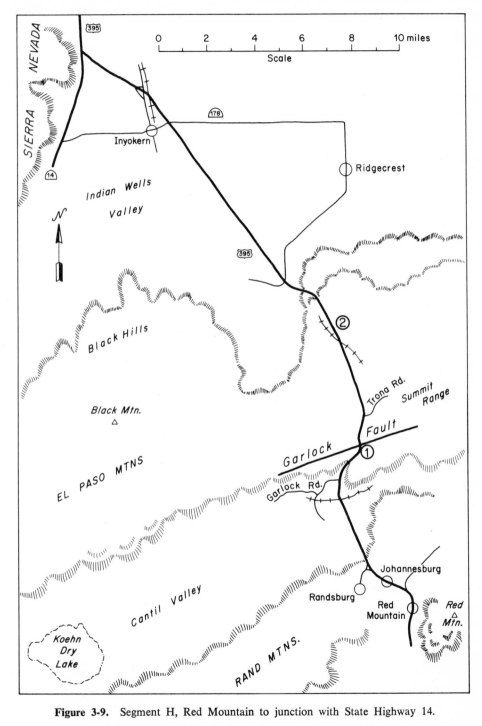

Figure 3-9. Segment H, Red Mountain to junction with State Highway 14.

U.S. Hwy. 395

Slate Range

So. End Death Valley

Searles Lake

Ridgecrest

Avawatz Mtns.

Garlock Fault

Eastern
Mojave
Desert

Photo 3-10. High-altitude oblique view looking east along Garlock fault. (U. S. Air Force photo taken for U. S. Geological Survey, 744L-026).

● About 8.5 miles down the straightaway we pass the junction with State Highway 178 and in less than 2 miles farther rise onto an overpass across the railroad. From here we get a good view of the gap at Little Lake (at 1:30 o'clock) into which we shortly head. It separates the Sierras from the Coso Range. Volcanics mantling the south flank of the Coso Range are seen at 2 o'clock.

● We proceed to a junction with State Highway 14. Now shift to Segment M to continue the trip north to Mammoth.

LA CAÑADA EXIT

*Segment I—La Cañada to Vincent Junction,
35 miles, Figure 3-10*

• This segment follows a winding highway across the San Gabriel Mountains. Alertness is required to spot some locations, and the geology is a little complex. Take the trip leisurely, make as many of the suggested stops as possible, read ahead in the trip guide, and do a little navigating by keeping track of check points such as bridges, passing lanes, powerline crossings, and mileages. Many people, including me, get car sick from reading on winding roads. You may be more comfortable doing your reading parked in one of the many turnouts. Things get easier beyond the crossing of Big Tujunga Creek. Some locations are numbered on the map (Figure 3-10) and correspondingly in the trip guide to facilitate correlation. Study the map and the trip guide before starting this journey to get a feeling of scale and the distribution of features.

• In La Cañada, turn north off Foothill Boulevard onto State Highway 2 (Angeles Crest). (Note your odometer reading.) We are headed due north up the alluvial slope at the foot of San Gabriel Mountains. At its head, this straight stretch is reputed to have the steepest grade of any part of the Angeles Crest-Angeles Forest Highway. Read ahead in the road guide.

• In one mile the highway curves east, we cross Gould Canyon on a concrete bridge and immediately enter road cuts in crystalline rocks of the San Gabriel Mountains block. In passing from the alluvial slope to the mountain block, we have crossed the Foothill fault zone which determines the south face of the mountains. Displacements up to 6 feet occurred on fractures in this zone near San Fernando in the 1971 earthquake.

• The igneous rock exposed in road cuts opposite the La Cañada Country Club turnoff (Starlight Crest Dr., Golden Helmet Restaurant) is part of a rock group called the Wilson diorite. The name comes from Mt. Wilson where the TV towers and telescopes are. The Wilson diorite is a member of a younger class of igneous intrusive bodies in the San Gabriels, about 80-90 m.y. old. Some 200 feet beyond the turnoff, a well defined 3-foot dike of white igneous rock cuts through the diorite.

• The road cut beyond the upper end of the golf course exposes first, old bouldery alluvial gravels, then gravels resting on crystalline rocks, and finally, just crystalline rocks.

• Rocks exposed in road cuts for the next mile are largely Wilson diorite. Just short of the Angeles Crest forest station the road enters a complex of old metamorphic gneisses and metamorphosed igneous rocks intruded by dikes and small irregular bodies of younger igneous rocks. This mixture is called the San Gabriel Formation. We pass through it for nearly 5 miles. All crystalline rocks in the San Gabriel Mountains have been severely deformed, badly fractured, and even shattered locally. Don't be disturbed if they look mixed up to you; even professional geologists regard them as challenging.

(1) Roughly 3 miles from La Cañada is the first of three large turnouts east of the highway. This one is the best for a stop. From it you get a good view into the narrow canyon of Arroyo Seco wherein flows a perennial stream, part of the Pasadena water supply. Downstream and also directly across the canyon are gently sloping topographic benches high on the canyon walls. These are remnants of former wide valley floors formed during pauses in the cutting of the Arroyo Seco. These pauses were terminated when renewed uplifting of the

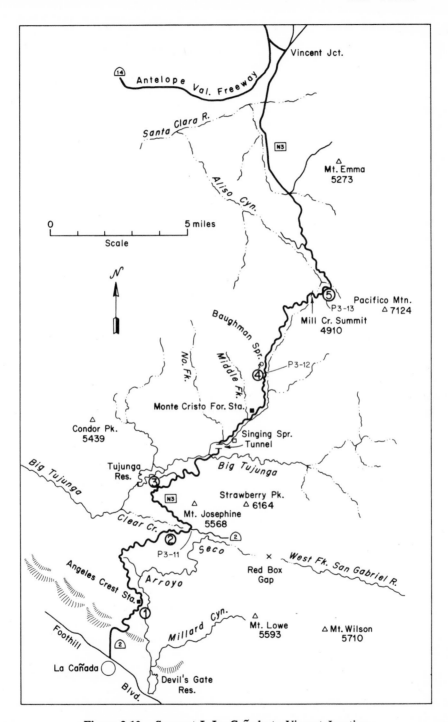

Figure 3-10. Segment I, La Cañada to Vincent Junction

mountains along the Foothill fault zone caused the stream to cut down again.

● We shortly pass the Angeles Crest forest station (on west side), and the mixed light, dark, and pink rocks seen in road cuts beyond are part of the San Gabriel Formation. At the first concrete bridge, about a mile beyond the guard station, pink dikes intruding dark rock should catch your attention. The dark rock is relatively rich in dark minerals containing considerable iron and magnesium, and the younger pink rock is relatively poor in these minerals and richer in those containing potassium, aluminum, and silicon.

● Within the next 2 miles the highway swings westward back into several canyons, crosses two more cement bridges, and then swings eastward out onto a narrow ridge which has a dirt road on its north side taking off southeasterly (6.3 miles from La Cañada). About ¼ mile farther, just beyond the Elevation 3000 Feet sign, are deep cuts on both sides of the highway exposing well banded gneiss cut by dikes.

● About a mile beyond the next passing lane (also a powerline crossing), note that the rocks become more uniform, lighter in color, and are characterized by rugged craggy slopes with large rock outcrops, as seen on hillsides to the east. We have crossed the several branches of the large Sierra Madre fault and are entering a body of igneous rock, the Lowe granodiorite, about 225 m.y. old.

② About 9 miles from La Cañada is the paved Angeles Forest Clear Creek vista. This is a good place to stop. If you do, you will see below the canyon of Clear Creek and on the north skyline Mt. Josephine with its lookout tower. The brown road cuts on the far side of Clear Creek are in the San Gabriel Formation. About

200-300 feet upslope is light colored rock which makes craggy outcrops extending to the top of the ridge. This is the Mt. Josephine granodiorite, only 80 m.y. old (compared to the 1700 m.y. age of parts of the San Gabriel Formation). The straight course of the contact between the light and dark rock, marked partly by green vegetation, suggests that it is a fault.

● Leaving the vista point, road cuts again expose dark rocks cut by younger pink dikes. In a mile is the junction of California Highway 2 (Angeles Crest) and County Highway N-3 (Angeles Forest Highway). We take the latter, but first pull off into the wide turnout on the right, opposite the intersection. On the skyline to the east is Red Box Gap where the road to Mt. Wilson turns off (Photo 3-11). The straight canyon running east toward Red Box is the head of the Arroyo Seco, which, to this point, has followed an irregularly winding course working generally northward. We are in a gap or saddle between the head of the Arroyo and Clear Creek, another straight canyon extending west into the Big Tujunga. It is aligned with the head of the Arroyo and the two gaps. If we were at Red Box we could see that the same alignment is extended eastward, first by the West Fork of San Gabriel river and then by the East Fork. This is no accident. All these features line up along the trace of one of southern California's major faults, the San Gabriel (Figure 1-1). Rapid erosion of ground-up rock along the fault zone has created this alignment of canyons and gaps.

● (Record your odometer reading before leaving the junction.) We now follow Angeles Forest Highway (County N-3) toward Palmdale and in 0.2 mile enter a section of four-lane highway. The first road cuts expose gravel succeeded by brown weathering

Photo 3-11. Looking east up head of Arroyo Seco to Red Box gap from junction of Angeles Crest and Angeles Forest highways.

rocks which are part of the San Gabriel Formation. Some subsequent cuts are again in gravel.

● Within a mile from the junction we approach the contact with the lighter-colored Mt. Josephine granodiorite and in another half mile are well into that body.

③ About 2.5 miles from the junction is a small (one-car size) turnout on the west side of the highway (under the power-

lines again), which gives a good view of Big Tujunga Canyon and the Big Tujunga flood-control dam and reservoir. (This stop is a little awkward; if traffic is heavy, don't attempt it.) To be effective, a flood-control reservoir should not be filled with water or rock debris. One of the problems with Big Tujunga is debris. It lost nearly 25 per cent of its capacity in a single year, 1938, through filling by sand and gravel.

• In another 0.2 to 0.3 mile, road cuts just beyond the Rock Slide sign expose old angular blocky rockslide or rock-fall debris, which, as the sign indicates, is prone to shed blocks onto the highway even today.

• We continue in the Mt. Josephine granodiorite for several miles. Locally it displays some darker phases, but some of the dark masses seen are clearly inclusions of older rock because they are shot through with dikes of Mt. Josephine granodiorite and display gneissic banding.

• About 6 miles from the junction we approach and cross the narrows of Big Tujunga Creek on a high-arch concrete bridge. The dark rock at both ends of the bridge is a phase of the Mt. Josephine intrusive body which is somewhat poorer in silica and richer in iron, calcium, and magnesium than the lighter phase. A stop at the turnout beyond the bridge and a walk back to the lookout over the gorge are worthwhile. This gives you a close view of the dark phase of the Mt. Josephine granodiorite and reveals some smoothly shaped inclusions, 3 inches to 2 feet across, of very dark older rock which it contains.

• Entering the tunnel 0.4 mile up the highway, we see exposures of strongly banded gneiss at its south portal. The gneiss is simply a large chunk of old rock (1700 m.y.) caught up in the Mt. Josephine granodiorite (80 m.y.). The north portal is again in Mt. Josephine granodiorite.

• As you pass Forest Springs go slowly and watch carefully the road-cut exposures beyond. Much of the rock is granular, disintegrates easily, and is brownish to gray. Some is mixed white and black, with blocky fracture, and is more solid looking. The disintegrating and solid types are interspersed, but by the time we pass the Monte Cristo Ranger Station (0.4 mile) only the white and black rock is seen. We have just crossed a zone in which Mt. Josephine granodiorite (disintegrating) intrudes a much older (1220 m.y.) rock known as *anorthosite*. Samples from the moon suggest that anorthosite may be a major constituent of the lunar uplands; so geologists are very interested in this San Gabriel anorthosite. We travel through the anorthosite body for the next 3.5 miles. Note the whiteness of the adjoining hill slopes.

Anorthosite is an unusual igneous rock made up almost exclusively of a calcium-rich feldspar mineral, which here forms crystals commonly several inches to occasionally 2 or 3 feet across. In many places this particular body of anorthosite displays a steeply inclined banding of dark and light layers. The darker layers comprise a different type of rock called *gabbro*. It contains some large (many inches across) crystals of a greenish mineral, *pyroxene*, and also a titanium-rich iron oxide mineral, *ilmenite*. In some larger exposures the gabbro is spotted. The light gray to white layers are the anorthosite, some fresh specimens of which have a delicate lavender color. This layering originated in an essentially horizontal position through crystallization and settling of minerals within a molten body. The solidified body and layers were later tilted to their present inclination.

Anorthosite masses are rare, and we are fortunate to have one in our own backyard. Because of the similarity to the moon rocks, current interest in it is at a high level. Many of the lunar rocks are abnormally rich in titanium, and sure enough, associated with our anorthosite body is a lot of ilmenite, a titanium-rich iron oxide.

④ We pass Monte Cristo campground (east) and in less than a mile enter a four-lane section where the dark rocks are mostly gabbros associated with the anorthosite. In 0.2 mile, where the highway curves east just short of Baughman Spring (the watering spot and cottonwood tree west of the highway), is a wide turnout on the west side. This is a good place to stop to see anorthosite close up. In the steep face west of the turnout, banding is unusually good (Photo 3-12); the white pointed structures are crystal growths indicating that the top of the original anorthosite mass is to the north. In the little gully at the north end of this cut, you can find chunks of black, heavy, titanium-rich ilmenite and large fragments of greenish pyroxene crystals.

● In less than a mile beyond Baughman Spring, exposures of dark well-banded rock are followed by strongly banded, lighter materials. This marks the contact between the anorthosite body and a large mass of Lowe granodiorite through which we now drive for the rest of our crossing of the mountains. The rocks at the contact are banded (gneissic) because of the high degree of shearing to which they have been subjected in this contact zone. The banding in the Lowe granodiorite dies out northward up the road.

Photo 3-12. Banding of anorthosite and gabbro layers near Baughman Spring, San Gabriel Mountains. White triangular areas at left are crystallization structures pointing to top of original magma body.

The road cuts, on up to Mill Creek Summit, give excellent exposures of the Lowe granodiorite. It is a little browner than the Mt. Josephine body and looks coarser grained. Note the typical craggy slopes west of the highway.

⑤ The Lowe granodiorite also contains some distinctive, dark, fine-grained sub-horizontal dikes (Photo 3-13). These are most apparent in road cuts going down the other side of Mill Creek Summit which we cross at 4910 feet elevation. (Note odometer reading at the summit.)

• About 2 miles beyond Mill Creek Summit is a good view at 11 o'clock of the Soledad Basin and upper Santa Clara River-Ac-

Photo 3-13. Dark subhorizontal dikes intruding the Lowe granodiorite as exposed in road cut north of Mill Creek Summit.

ton area. In another 2 miles, just beyond Aliso Canyon road, gneissic banding is again seen in the Lowe granodiorite.

● Five miles from Mill Creek Summit we emerge onto a straightway, with a passing lane, descending a gently inclined alluvial slope. Many of the road cuts beyond the straightaway expose alluvium through which knobs of rock occasionally project.

● The Edison Company distribution station sits on top of a westward sloping, well-graded, alluvial surface. The material in the road cut, just beyond it, is deeply weathered crystalline rocks.

● Soon we emerge into the open gap at the head of the Santa Clara River drainage through which the railroad and freeway find access to the desert. This is Soledad Pass, and its size and openness suggest that at earlier times considerable drainage may have passed this way, perhaps from the desert to the sea. Currently, the headwaters of Santa Clara River are eating back into the gap, and eventually they may work through to the desert.

At 35 miles from Foothill Boulevard we come to Vincent Junction and with a zig and a zag join the Antelope Freeway (State Highway 14). If continuing to Mammoth, skip to Segment K; if returning to Los Angeles, run Segment J in reverse, from end to beginning.

SAN FERNANDO EXIT

Segment J—San Fernando to Vincent Junction (Palmdale), 38 miles, Figure 3-11

● This exit from the Los Angeles Basin for the Mammoth trip begins at the intersection of Interstate highways 5 (Golden State Freeway) and 405 (San Diego Freeway) near Van Norman Reservoir at the northwest corner of San Fernando Valley. It follows State Highway 14 (Sierra Highway and Antelope Valley Freeway) to a junction with Segment I at Vincent Junction near Palmdale.

● In approaching the starting point by way of either freeway, you traverse San Fernando Valley, a large east-west synclinal downwarp. It is bounded by the crystalline-rock, fault-block masses of the Verdugo Hills and San Gabriel Mountains on the north and by anticlinally upwarped Santa Monica Mountains on the south. The Santa Monicas consist chiefly of late-middle to early Tertiary (12-70 m.y.) sedimentary rocks, largely marine, with some Triassic slates (200 m.y.) and granitic rocks (100-150 m.y.?) to the east and mid-Tertiary (20-25 m.y.) volcanic rocks to the west. San Fernando Valley contains nearly 1500 feet of alluvial gravels resting upon thousands of feet of mid-Tertiary marine sedimentary beds and volcanics. These rocks are underlaid by still older materials laid down over the region before downwarping of the valley was initiated.

● Approaching the junction of highways 5 and 405, one sees the high, rugged, crystalline-rock San Gabriel Mountains to the north and the Santa Susana Mountains, largely marine mid- and late-Tertiary sedimentary rocks, to the northwest. If the day is clear, you should be able to make out the rough, rugged skyline of early Tertiary (60-70 m.y.) and late Mesozoic (80 m.y.) massive sandstone outcrops in the Simi Hills closing the west end of San Fernando Valley.

● Both freeways enter the low Mission Hills about a mile before juncture, and initial road cuts on each expose steeply tilted, late-Miocene (15 m.y.) marine shales. The Miocene exposures are small on the Golden State but extensive along the San Diego Freeway where some folding within the beds is recognizable. Succeeding the Miocene are marine sandstone, siltstone, and conglomerate beds of early to late Pliocene (4-12 m.y.) age with more gentle inclinations. Beyond the junction, a good section of gently dipping late Pliocene deposits (2-4 m.y.), consisting partly of marine sandstone, siltstone, and conglomerate layers, is visible along the far shore of Van Norman Reservoir. Northwestward these beds grade upward into younger and coarser sandstone and conglomerate deposits of the Pleistocene (2 m.y.) Saugus Formation.

● Ahead at the aqueduct-outfall cascades, on the hillside at 12:15 o'clock, is a complex structural situation. The irregular Foothill fault zone which bounds the south base of San Gabriel Mountains to the east meets here the Santa Susana thrust fault which has curved northward from its position about half-way up on the south face of Santa Susana Mountains to the west. Here the thrust lies about at the base of the brown cliffs north of the freeway, just beyond the cascades. The situation is complicated by several other small, cross-cutting faults extending northeasterly. It is not possible to tell whether the Foothill fault truncates the Santa Susana, or vice versa, or whether one fault grades into the other. Both lateral and vertical displacements of as much as 6 feet occurred on members of the

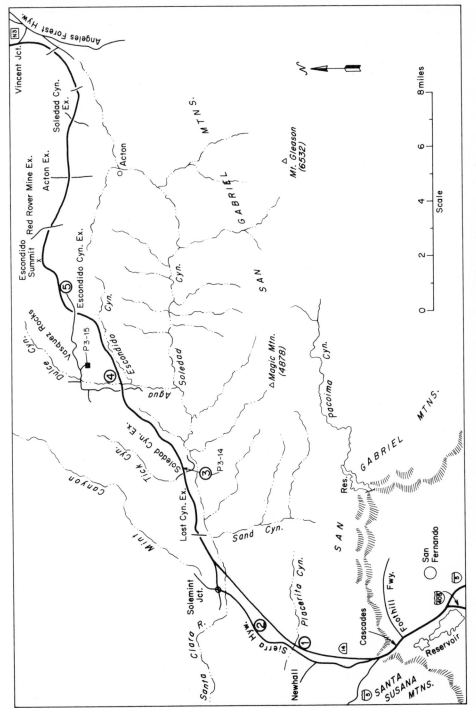

Figure 3-11. Segment J, San Fernando to Vincent Junction.

complex Foothill fault zone a few miles east of here during the San Fernando earthquake of February 9, 1971.

• The cliffs beyond the cascades and the rocks exposed in the huge road cuts of the "bucket of worms" freeway interchange ahead are marine sedimentary beds of late Pliocene age (3-5 m.y.). There is much fine sandstone, siltstone, some shale, and considerable conglomerate. Some of the exposures show remarkable examples of scour channels, which, filled with coarse debris, cut into underlying beds. The tilt of these beds differs greatly in direction and steepness from place to place, an expression of the locally complex structure.

Damage from the San Fernando earthquake of 1971 has caused delay in completion of the Antelope Valley Freeway between here and the Santa Clara River. Guide notes for the next 7 miles are designed to serve both the old (Sierra Highway) route of State Highway 14 and the new freeway, when completed.

• As you turn off on either old State Highway 14 or the Antelope Valley Freeway to Palmdale, note your odometer reading. The road cuts, beyond the interchange for a little more than 2 miles, are in early Pliocene to late Miocene (8-15 m.y.) marine sedimentary beds of a character similar to those seen at the interchange. Through here we pass some old oil fields and refineries dating back to 1889.

① In a little over 2 miles from the interchange we pass the turnoff to Newhall, cross Whitney Canyon, and enter road cuts in nearly horizontal, brown, coarse, nonmarine conglomerate and sandstone beds of the Saugus Formation, Pleistocene in age (2 m.y.). Cuts are particularly spectacular on the east side of the new freeway coming down into Placerita Canyon.

• In another mile we cross Placerita Canyon, so named for the discovery of placer gold in 1842 by the Spanish, 6 years before the find at Sutter's mill which touched off the great Gold Rush of 1849.

• Beyond Placerita Canyon are more of the gently dipping conglomeratic Pleistocene beds. Here we are passing the abandoned tanks and wells of the almost depleted Placerita oil field, a forlorn and depressing testimony to improper exploitation of a natural resource. This field, discovered in 1948, proved to be a prolific producer from relatively shallow depths. However, the area had once been subdivided by a land promoter, and town-lot plats were held by different owners. Everyone got into the act, and too many wells were drilled in the rush to get the oil out of the ground, with the result that only a part of the full potential of the field was realized.

② If you are traveling the old Sierra Highway, watch road cuts on its west side as you approach the top of the ridge a little over a mile beyond Placerita Canyon. The heretofore nearly horizontal bedding steepens suddenly and then flattens out again on the other side of the ridge crest. You have just crossed the San Gabriel fault (Figure 1-1), a major southern California fracture with many miles of right-lateral displacement. The Placerita oil field owes its existence in part to the trapping of oil in beds upturned along this fault. At this writing it is not possible to tell whether cuts along the new freeway will expose the near-vertical beds marking the trace of this fault.

• As we descend from the crest of the Placerita ridge, the broad upper Santa Clara River Valley opens out ahead. We shortly dip down and cross the sandy (usually dry) bed of the river itself. In Solemint Junction turn right, and in about ¾ mile you enter the completed section of Antelope Valley Freeway. When the freeway is fully completed, Solemint will be bypassed.

● In the large road cut on the west side of the freeway just beyond Sand Canyon Road exit is a good exposure of gently inclined, light-colored sandstone and conglomerate beds of the terrestrial Upper Miocene (12-15 m.y.) Mint Canyon Formation. Outcrops on hillsides north of the freeway in the next 2 miles reflect the gentle inclination of resistant beds in this formation. We are now in a local Tertiary basin of deposition on the northwest flank of the San Gabriel Mountains, known to geologists as Soledad Basin. Some 25,000 feet of land-laid deposits accumulated here, part of them remarkably

coarse. We pass through many thousand feet of inclined Mint Canyon beds before getting to a still older deposit. With minor exceptions, the road cuts ahead for the next 6.5 miles are in Mint Canyon beds.

● Approaching the Soledad Canyon exit an extensive sand and gravel operation lies in the bed of Santa Clara River just east of the freeway. In 1970 the total value of sand and gravel marketed in California exceeded 160 million dollars. It was the fourth most valuable natural-resource product of the state after oil, gas, and cement.

③ Just beyond the Soledad Canyon overpass in a large cut on the west side is a striking angular unconformity (Photo 3-14) between nearly horizontal, iron stained,

Photo 3-14. Angular unconformity between gently tilted beds of late Miocene Mint Canyon Formation and horizontal Pleistocene gravels in road cut at Soledad Canyon Road junction on Antelope Valley Freeway.

Pleistocene (1-2 m.y.) gravels resting upon light colored, tilted, non-marine sandstone and conglomerate beds of the mid-Tertiary (12-15 m.y.) Mint Canyon Formation. This relationship can best be viewed by taking the off-ramp and then driving slowly up the on-ramp to get back on the freeway.

● In the next road cut on the west side, 0.3 mile north, is good exposure of the iron stained gravels. Just beyond this point, about a mile away at 2 o'clock, one gets a view of the mouth of Soledad Canyon. The rocks there are crystalline igneous materials. Soledad Canyon discharges the drainage of most of the northwest flank of the San Gabriel Mountains, which make up the high skyline to southeast. We are traveling a course which has already circled their west end.

● For the next 3.5 miles the road cuts and intervening hillsides provide good exposures of Mint Canyon beds, all tilted at a modest angle. Some of the more massive, better-cemented sandstone and conglomerate layers make cliffs, bluffs, and ridges of various colors and characteristics on the hillsides.

These beds, along with the similar but locally coarser and more reddish layers of the underlying Vasquez Formation, constitute a deposit exceeding 20,000 feet in thickness laid down rapidly within the Soledad Basin. At times the ocean lay along the western edge of this basin but never penetrated very far into it. The adjoining country, particularly to the southeast, was high standing, and it shed great quantities of very coarse, broken-up rock into the basin in a succession of episodic events. At other times conditions were quieter, and finer sandy and silty beds were laid down along stream courses and in shallow ponds. The major part of the sedimentary deposition was preceded by extrusion of several thousand feet of lavas, which we see shortly. The

Mint Canyon beds have yielded remains of fossil horses, camels, and rodents, but no vertebrate fossils have been found in the Vasquez Formation.

● At Aqua Dulce Canyon one can exit and take a short drive north to Vasquez Rocks, a county park in striking and picturesque hogback ridges formed by erosion of resistant, red sandstone and fanglomerate layers in the Vasquez Formation (Photo 3-15).

④ In the first road cut on the north side just beyond the Agua Dulce intersection, some grayish-brown, very coarse, bouldery fanglomerate layers rest upon pink fine siltstone and sandstone layers. This is the contact of the Mint Canyon resting on the older, Vasquez Formation. The Vasquez beds shortly become much coarser. Note the cavernous weathering in sandstone ledges on the hillslopes beyond this exposure.

● If you look northwest at 9 o'clock through a gap 1.6 miles beyond Agua Dulce Canyon, you can catch a quick glimpse of Vasquez Rocks.

● In another half mile dead ahead are dark volcanic hills. These are part of a thick section (4500 feet) of successive lava flows within the lower Vasquez Formation.

● Two and one-half miles beyond Agua Dulce Canyon a deep road cut gives excellent exposures of the lavas. The immediately succeeding road cuts and the intervening slopes are also in lava.

● Just beyond the Escondido Canyon exit the brown ridge to the northwest at 9 o'clock, with the airway beacon on top, is underlaid by a granite-like rock called syenite which has been dated as 1220 m.y. old (Precambrian). To the east at 2 o'clock on the skyline, dipping fanglomerate beds are seen in the Vasquez Formation underlying the lavas.

Photo 3-15. Vasquez Rocks formed by differential erosion of sandstone and fanglomerate beds in early mid-Tertiary Vasquez Formation north of Antelope Valley Freeway.

⑤ One mile from Escondido Canyon exit, note the whitish area on the lower hill slopes just southeast of the freeway where considerable grading has been done. It is underlaid by fanglomerate beds containing huge fragments of an unusual igneous rock called anorthosite, which is exposed as bedrock in place within the San Gabriel Mountains to the southeast. Anorthosite is a rare rock made up largely of large crystals of calcium-rich feldspar. It is of particular current interest because a major component of the uplands of the moon may be anorthosite. Extensive exposures of the rock are seen along Mill Creek on the Segment I field trip, where it is more completely described.

● Just beyond Escondido Summit (3258 ft.), the Truck Parking Area is a good place to stop. The rocks here are Precambrian igneous and metamorphic types from which a small production of gold has been obtained in times past. The high skyline peak at 12:15 o'clock (snow covered in winter) is Mt. Williamson (8214 ft.). The skyline ridge at 9 o'clock is Sierra Pelona. The open valley ahead is partly filled with alluvium and represents an older landscape which is about to be dissected by tributaries of Soledad Canyon eating headward into it.

● One mile beyond the Red Rover Mine road we get a view of other high peaks within the San Gabriels, Pacifico Mountain (7124 ft.) at 1 o'clock and Mt. Gleason (6502 ft.) at 2 o'clock. Most of the front of the San Gabriels as viewed from here is made up of Lowe granodiorite, an intrusive igneous rock about 220 m.y. old, which is extensively seen on the Segment I trip.

● About 0.7 mile beyond Crown Valley Road exit, road cuts on both sides of the freeway expose more lavas, part of a down-dropped fault block.

● One-half mile beyond the Soledad Canyon Road exit is a huge road cut exposing rocks which belong to the Lowe granodiorite group; here they are badly fractured and deformed.

● Soon we come to the Pearblossom-Angeles Forest exit and shortly thereafter effect a junction with field trip Segment I which has come across the mountains from La Cañada. Proceed north now on Segment K, or return to Los Angeles via Segment I.

Segment K—Vincent Junction (Palmdale) to Mojave, 40 miles, Figure 3-12

● Beyond the junction with Segment I, Antelope Valley Freeway passes through five successive ridges in deep notches with cuts on both sides. Notches 1, 3, and 4 expose tilted layers of badly altered and weathered lavas, or lavas and volcanic agglomerates. Notch 2 exposes fanglomerates and agglomerates, and notch 5 is in old crystalline rocks.

● The vista point at 2.5 miles from Vincent Junction is worth a stop. Here one sees the western Mojave Desert, Palmdale Reservoir, California Aqueduct, and the San Andreas rift. On the skyline at 11-12 o'clock are Tehachapi Mountains, and on the 3 o'clock skyline are typical buttes and knobs of the western Mojave.

Palmdale Reservoir lies in a vale created by erosion of the soft ground-up rocks along the San Andreas rift, here about one mile wide. This vale continues northwestward as Leona and Anaverde valleys. The most recent line of displacement, the break of 1857, lies along the far side of the reservoir and crosses the freeway just this side of the ridge through which the freeway passes in a deep cut. The ridge is a slice within the fault zone.

At the north end of the vista-point parking is a bronze plaque describing some of the historically significant earthquakes generated by the San Andreas. The last major displacement through here was in 1857, not 1957 as erroneously stated on the plaque. It occurred during the Fort Tejon earthquake and locally approached 20 feet in a right-lateral sense. The old (1913) Los Angeles-Owens Valley Aqueduct crosses the San Andreas fault underground near Elizabeth Lake, 15 miles northwest, but the new California Aqueduct crosses on the surface so that repairs can be made in case of a fault displacement.

● Back on the freeway we cross the aqueduct. Then 0.3 mile beyond the next exit (Avenue S) you should feel some bumps in the freeway, especially in the inboard lane. These are along the trace of the most recent fault break, and when they first developed shortly after the freeway was completed, thoughts were entertained that they might be the result of current creep on the fault. Subsequently, more bumps have formed within the notch just ahead, and all the bumps may be simply the result of differential settlement of macerated rocks within the fault zone.

① In the cuts of the notch, note the severe folding and crumpling of shale and siltstone beds within the mid-Pliocene (5-8 m.y.) Anaverde Formation (Photo 3-16).

● As the freeway clears the low hills, we enter Antelope Valley, a flat, largely alluvial area. On the eastern skyline are low residual bedrock peaks, ridges, and rounded domes greatly reduced by erosion. Westward the country is underlaid primarily by an unconsolidated alluvial fill mostly 1000-2000 feet thick but locally exceeding 5000 feet as shown by deep wells.

② A little north of Palmdale (just beyond Avenue P) look back at 7-9 o'clock to see the mountain front along the southwest edge of the desert as defined by branches of the San Andreas fault. The forelying Portal Ridge (Photo 2-3) at 9 o'clock is a fault slice on this side of the San Andreas. Lighter colored Ritter Ridge at 7 o'clock is of similar origin. The high skyline ridge behind at 7-8 o'clock is Sierra Pelona. It is composed of old metamorphic rocks appropriately named the Pelona schist, better known to some of you as Bouquet

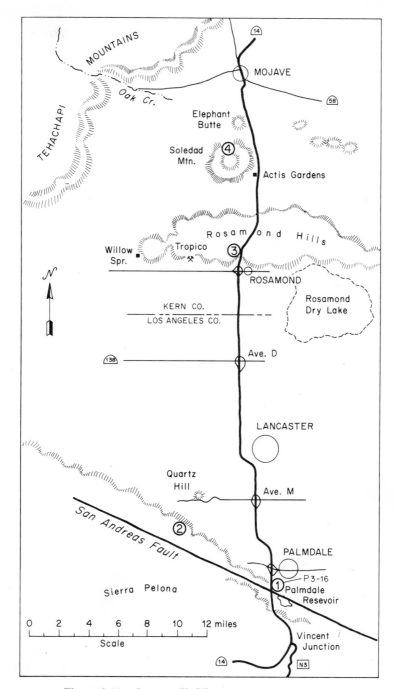

Figure 3-12. Segment K, Vincent Junction to Mojave.

Photo 3-16. Crumpled beds of Pliocene Anaverde Formation in fault-slice ridge within San Andreas fault zone along Antelope Valley Freeway near Palmdale.

stone from quarries in Bouquet Canyon. This is one of the more widely used southern California building stones.

● This is a good stretch in which to relax. We see little geology close at hand until the highway climbs into hills near Rosamond. If the weather is clear, just north of Lancaster you get a sense of the wedge shape of the western Mojave as defined by convergence of structures related to the San Andreas fault zone behind and the Garlock fault zone ahead (Photo 2-3). The point of the wedge is 30 miles west at about 9:30 o'clock. Tehachapi Mountains make up the far northwest skyline, and the high peak at about 10:30 o'clock is Double Mountain (7,988 ft.). The part of the Tehachapi seen from here is composed primarily of granitic rocks inclosing masses of older metamorphics which contain marble beds that furnish carbonate rock for the large cement plants of the Tehachapi-Mojave area.

● (Note your odometer reading at the Kern County line (Avenue A).) From here the Rosamond Hills are seen dead ahead. Their south margin is defined by the Rosamond-Willow Springs fault. Here the hills are composed of granitic rocks with a mantle on the south flank of younger deposits of the mid-Tertiary (15-30 m.y.) Tropico

Group named from Tropico Mine about 4 miles west. East of Rosamond the Tropico Group contains considerable fine-grained lake deposits, but here it consists largely of deformed layers of fine fragmental volcanic debris (tuff beds), sandstone rich in volcanic particles, volcanic fanglomerates, and lavas with related near-surface intrusive bodies such as plugs, pods, and dikes.

③ In the hills just ahead, the smoother white to light green slopes are underlaid by fine-grained volcanic tuff and tuff-breccia beds. The darker craggy outcrops are lavas, shallow intrusive bodies, or granular fanglomerate. In the first road cuts and on adjacent slopes you will see rocks of a variety of colors, white, cream, green, red, and brown in various shades. These rocks have been locally quarried for roofing granules, and such a combination of colors usually indicates the presence of considerable fragmental volcanic debris.

About 30 miles east-northeast in the vicinity of Boron, the Tropico Group contains the world's richest known deposits of borate minerals (*see* Mojave Desert province, resources, and field trip Segment G). One of the abundant minerals is kernite, named in recognition of this famous locality in Kern County.

④ About 7 miles beyond the county line the highway straightens out and Soledad Mountain (4183 ft.) looms up at 10:30 o'clock. In times past, Soledad Mountain and Elephant Butte (the smaller mountain northeast of it which you will see shortly) were the sites of productive gold and silver mines, the Golden Queen and the Silver Queen being two of the more famous. You will see many abandoned mine workings on the lower slopes of these mountains, the waste piles of the Golden Queen being the most prominent on the north slope of Soledad Mountain. Total production from the district amounted to roughly 26 million dollars.

Soledad Mountain is a cluster of intrusive plugs, pods, and dikes of fine-grained, reddish-brown igneous rocks called rhyolite, an example of which is exposed in the cut on the highway about 1 mile beyond the trees and houses at Actis Gardens. These bodies were intruded into older granitic rocks, probably in Miocene (12-25 m.y.) time, and the ore mineralization is thought to be related to this intrusive episode. Fine-grained intrusions of this type are usually emplaced near the surface where they cool quickly. The associated ore deposits are sometimes fabulously rich but normally do not extend to great depth. Mining here reached a maximum depth of 1000 feet.

● The low knobs and ridges east of the highway opposite Soledad Mountain are, in large part, similar—small satellite intrusive plugs etched out by erosion.

● Beyond Elephant Butte, look west at 9:15 o'clock in passing Purdy Road to see, 7 miles off, the dust from a large cement plant and the light colored areas on the hills nearby. These light areas are the quarries from which the marble for cement is obtained.

● Coming into Mojave the lower more subdued skyline at 10-11 o'clock indicates the location of Tehachapi Valley (Photo 3-17) which separates the Tehachapi and southern Sierra Nevada mountains.

Segment L—Mojave to Junction with U. S. 395, 49 miles, Figure 3-13

● We leave Mojave traveling north-north-east on State Highway 14 toward Bishop. ▶ (Note odometer reading at the **Y** at the north edge of town.) Our route to Redrock Canyon parallels the south end of the Sierra Nevada as cut off abruptly by the Garlock fault zone (Photo 3-17).

① Looking at 9:15 o'clock beyond the first curve, about 2.7 miles from town, you see a flat topped ridge with a steep south face in front of the main mountain mass. It is a little forelying fault block (Photo 3-18) lifted up along one of the many parallel fractures constituting the Garlock fault zone. Back to the west the railroad and highway from Mojave to Tehachapi turn west-southwest just within the mountains, following a route up Cache Creek (8 o'clock) which has carved out its canyon along the main trace of the Garlock (Photo 3-18).

Also from here, and at many other points along the way to Redrock Canyon, good views are seen at 2-3 o'clock of the string of buttes extending east from Mojave. The large one at 2 o'clock is Castle Butte. These prominences are mostly plugs of fine-grained intrusive rocks, etched out by erosion, or knobs capped by a protective sheet of lava. The broad smooth surfaces between are either alluvium or masses of granitic rock worn down by weathering and erosion. In clear weather with low-angle lighting this combination makes a striking landscape.

● About 5.5 miles out is the Randsburg Cutoff-California City intersection. Along the straight stretch of highway beyond, we head directly toward El Paso Mountains on the skyline. The skyline hills at 1:30 o'clock are the Rand Mountains, at the eastern end of which are the famed mining camps of Randsburg and Johannesburg (*see* field trip Segment G).

The relatively smooth surface across which we travel is an alluvial apron made up of a series of coalescing fans spreading outward from the base of the southern Sierra Nevada. The fans are smooth and gently sloping because they are composed of fine-grained debris, mostly disintegrated granite. You see only occasional cobbles and pebbles. The southern Sierras are made up principally of granitic rocks with some thin slices of metamorphics along the range front giving the variegated colors. Tertiary sedimentary and fragmental volcanic materials crop out farther back in range.

● The next 8 or 9 miles are uneventful; so we insert at this time some information on the Ricardo Formation, extensively exposed in Redrock Canyon ahead (Photo 3-19). There is so much to see at Redrock that you need a little preparation.

The Ricardo is a terrestrial (land-laid) deposit nearly 7000 feet thick that accumulated in a local land locked basin during early Pliocene time (10 m.y.). The formation consists primarily of layers of sand, silt, clay, and gravel, all containing considerable fragmental volcanic debris. The highly variegated colors of the lower part of the Ricardo reflect this volcanic content. Also interbedded are two groups of dark lava flows which cap prominent ridges etched out by erosion. Notable near the center of the Redrock amphitheater is a thick, massive, delicate pink bed of volcanic tuff-breccia formed by eruption of hot volcanic ash and other fragmental volcanic debris.

The Ricardo deposits are soft; hence they are eroded easily and rapidly. Yet some of the layers are coherent enough to stand in vertical faces. This property of easy erosion, coupled with a contrast in the coherence and resistance of different layers, helps create

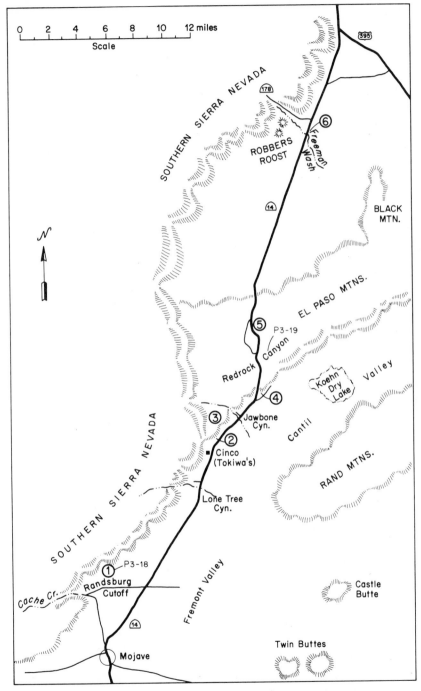

Figure 3-13. Segment L, Mojave to U. S. 395 junction.

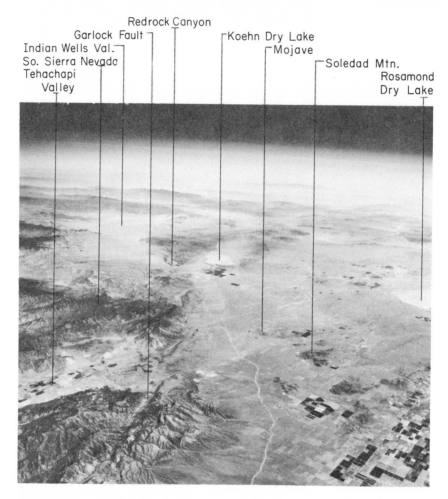

Photo 3-17. High-altitude oblique view looking northeast along trace of Garlock Fault. (U. S. Air Force photo taken for U. S. Geological Survey, 064L-093).

the castellated forms, spires, chutes, niches, alcoves, and other badland features which, along with the variations in color, give Redrock Canyon its striking character. A modest northwestward inclination of the bedding adds variety to the topographic forms. Most of you have probably seen Redrock Canyon in the background of a good many western thrillers, since this colorful and unusual terrain long ago caught the fancy of movie and TV companies.

The Ricardo beds are fossiliferous, and petrified wood is fairly common in places. Harder to find are the bones and teeth of vertebrate animals that inhabited the Ricordo basin 10 million years ago. They made up an interesting assemblage and included such animals as early horses, camels, mastodons, rhinoceroses, a variety of wild dogs, pronghorn antelope, deer, saber-tooth tigers, smaller cats, weasels, rabbits, and a goose. These animals lived in an open forest-grass-

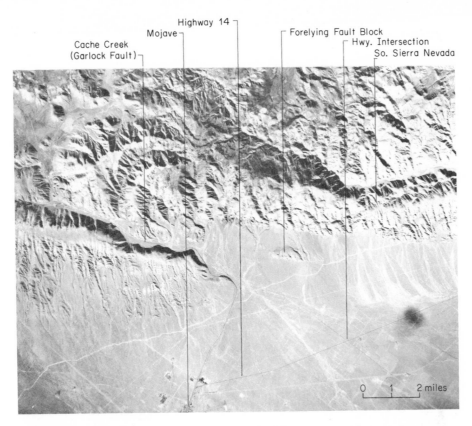

Cache Creek
(Garlock Fault)

Mojave

Highway 14

Forelying Fault Block

Hwy. Intersection
So. Sierra Nevada

0   1   2 miles

Photo 3-18. High-altitude vertical air photo of Mojave area, north toward upper right corner, scale at lower right. (U. S. Air Force photo taken for U. S. Geological Survey, 744V-031).

land environment with perhaps 15 inches of annual rainfall, roughly 3 times the present quota. Among the trees were live oaks, piñon pine, locust, cypress, acacia, and palms.

● Near Tokiwa's (Cinco), about 14 miles from Mojave, the highway curves eastward and from here to the mouth of Redrock Canyon we travel close along the mountain front. These mountains are bounded by El Paso fault, one of the principal fractures of the Garlock zone.

② In less than a mile beyond Cinco, a discerning observer can begin to see the fault from the highway where it crosses little gullies and spurs at the mountain front. The forelying spurs are lighter in color, softer-looking, and in places show layering. The steep, darker (locally lavender) mountain front behind is made up of highly fractured, fine-grained, granitic rock.

③ In another mile, red, green, brown, and purple rocks compose the mountain front. These are part of the Ricardo Formation, and their sudden appearance here is due to a north trending fault which drops them down against the older rocks on the west. This structure is considered by some geologists to mark the southern end of the Sierra Nevada frontal fault system, here truncated by El Paso fault. The hills east of the Sierra fault are lower and less rugged, being underlaid by softer rocks.

● A look left at 8 and 9 o'clock, about where the four-lane highway begins, shows Jawbone Canyon being crossed by both the new, 1970 (white), and old, 1913 (black), Los Angeles Aqueduct lines.

● In another mile the road to Randsburg branches east. In the next mile or two, approaching the mouth of Redrock Canyon, we pass through slightly dissected, light-tan, fine-grained, gently tilted Ricardo beds locally capped by younger stony gravels. Two miles beyond the Randsburg turnoff at 10 o'clock, the mountain front displays good development of badland topography in highly colored Ricardo beds. The black band is one of the groups of lava flows interbedded within the Ricardo.

④ Just before entering the narrow mouth of Redrock Canyon a quick look east at 2 o'clock shows a sharp contact between tilted Ricardo beds and crystalline rocks of the mountain front. This is El Paso fault again.

● After crossing the fault we enter the narrow, steep-walled gorge of lower Redrock Canyon which is cut into hard crystalline materials. It contrasts markedly with the broad amphitheater carved into much softer Ricardo beds, which will open out suddenly before us in less than a mile as we cross the depositional contact between the Ricardo deposits and the older crystalline rocks.

● East of the highway within the amphitheater is a spectacular castellated and fluted ridge (Photo 3-19) around the west end of which the highway curves. Approaching this curve you should see the massive, pink, volcanic tuff-breccia bed.

● To the west is another castellated ridge capped by dark lavas. We cross these lavas shortly ahead, but before that, note the wind-blown sand heaped on hillslopes west of the highway.

If you wish a more leisurely trip through Redrock, turn off west just beyond the pink tuff bed, and follow the old two-lane paved road to a junction with Highway 14 about 2.7 miles ahead.

⑤ As we climb out of Redrock Canyon on Highway 14, everyone except the driver can look back to 8 o'clock at the Elevation 3000 Feet sign and see the tilted Ricardo beds sharply truncated by a smooth erosion surface sloping gently eastward. This surface is what geologists call a pediment. It was cut by streams flowing southeastward

Photo 3-19. Gently tilted beds of Pliocene Ricardo Formation in Redrock Canyon.

out of the higher mountains to the west. Subsequently, the entire El Paso Range was uplifted and dissected to create the present topography, the pediment being partly destroyed in the process. Southbound travelers get a good view of these relationships just before dropping into Redrock Canyon.

● About 4 miles beyond the Redrock amphitheater we start another section of four-lane highway, only 1.5 miles long. As we top out at its end, a long straight stretch of highway extends ahead. The next curve is just over the distant skyline. Why not have everyone in the car guess how far it is to that curve? (Note your odometer reading as you leave the four-lane section.)

● As we move out onto the straightaway, at 2 o'clock is Black Mountain, so named for an obvious reason. The black rocks are basaltic lava flows whose normally dark color is accentuated by a coat of desert varnish.

● The long straightaway traverses an alluvial apron composed largely of disintegrated granite derived from the Sierras. Note the sparsity of boulders. In about 3 miles we begin crossing a succession of widely separated, shallow, flat-floored gullies cut into this apron. The dissection of any alluvial surface is an indication of some change in regiment of the streams flowing across it. In this instance, gentle uplift, or perhaps a change in climate has caused the shift from stream deposition to stream dissection.

The southern Sierra Nevada on the west, much lower and less rugged than farther north, are composed of a variety of igneous rocks with some metamorphic pendants. For the most part, these rocks weather easily; so the range front is subdued.

⑥ In another 7-8 miles we cross a deeper and larger gully identified as Freeman Wash by the sign on the bridge. Note the bouldery layers in its banks 100 feet

east of the highway and the large boulders on its floor. This unusually coarse material came down from the mountains as a mass of muddy flowing debris, something like freshly mixed concrete. As we rise up out of the wash look west at 8:45 o'clock to the mountain front where some forelying isolated ragged knobs, locally known as "Robbers Roost" claim attention. These are not something "pushed up out of the ground," as my mother told me as a child, but rather residual knobs left as the mountains, of which they were once a part, retreated westward under the attack of weathering and erosion.

● At and beyond Isabella turnoff we get good views, in decent weather, of Indian Wells Valley ahead (see Figure 3-14.) The mountains on the far side to the northeast are the Argus Range, the visible part being largely granitic. Their lower slopes are lightened in color by a mantle of wind-blown sand and silt swept up from the valley. Between 10,000 and 20,000 years ago the eastern and central parts of Indian Wells Valley held a shallow lake which eventually merged with a larger and deeper body of water in Searles Lake basin to the east (see Basin Ranges province, special features, Figure 2-14).

● Note your odometer at the first curve beyond Isabella junction and see how closely you guessed the length of the straightaway.

● We pass Homestead and Indian Wells, climb another little hill, and as the highway straightens out see far ahead at 10 o'clock volcanic cones and dark lava flows on the south end of the Coso Range behind Little Lake (Figure 3-14). These continue in view as we drive north.

● Shortly we intersect U. S. 395 coming in from the east and join those who elected to travel the San Bernardino-Cajon Pass route via Segments A, G, and H.

Segment M—Junction with U. S. 395 to Olancha, 41 miles, Figure 3-14

● Driving north from the junction of highways 395 and 14 keep an eye on the Coso Range at about 12-1 o'clock ahead. The volcanic cones and sheets of dark lava thereon become more apparent on closer approach.

● The front of the Sierra Nevada to the west becomes higher and steeper northward to Lone Pine. Here it is still somewhat subdued, partly because the rocks composing the range front are older granitics with some metamorphic pendants which weather readily so that the slopes are largely debris mantled rather than bare rock. A half mile north of Green Acres (1.7 miles from junction), a good view is seen at 9:15 o'clock up Grapevine Canyon to Owens Peak in the core of the range where hard, clean, fresh granitic rocks have a more rugged topographic expression resembling the range front farther north.

▶ ● Check your odometer at the Inyo County line as we enter a stretch of four-lane highway and in about 3.5 miles pass a lumber mill.

① About 1.5 miles beyond the mill and 150 yards west are some low cliffs and cuts along the other lanes of the separated highway. Even from here you can see that some of the gravel layers exposed in the cuts are exceptionally rich in large boulders. This coarse material has been carried from the mountains as sheets of muddy, bouldery material called debris flows. The carrying power of such flows is great, and many of the huge boulders on alluvial fans farther north in Owens Valley have been transported by this means. Debris flows have occurred repeatedly in the past and currently take place following cloudburst rains in the mountains (Photo 3-20).

● Beyond the railroad crossing, cliffs of black lava lie to the east. Ahead is a narrow gap through which the highway, railroad, and aqueduct lines pass (Photo 3-21). The gap and the lava cliffs have been carved by a large stream that once flowed through here carrying meltwaters from ice age glaciers of the Sierras. We shall call this the glacial Owens River. It flowed south and eventually east to feed a chain of large lakes in basins as far east as Death Valley (*see* Basin Range province, special features, Figure 2-14).

● Record odometer reading opposite the ◀ Little Lake Hotel. Just beyond we come to the lake itself, a shallow body occupying part of the old river course. The lake is nourished by small seepage springs.

● Passing the lake, Red Hill, a cone of reddish volcanic cinders looms ahead at 12 o'clock. The lava cliff on the east, north from the lake, displays a prominent columnar structure. This is caused by fractures formed as the lava cooled, and is known as columnar jointing. It is superbly developed at Devil's Postpile on the Middle Fork of San Joaquin River west of Mammoth. Why the Devil gets credit for creating columnar joints remains a mystery.

● About 1.5 miles north of Little Lake Hotel at the curve in the highway, look quickly northeast at 1:30 o'clock for a view of an abandoned gorge cut into the lavas by glacial Owen River. At its head is a spectacular dry waterfall with deep pot holes and stream-polished rock surfaces, which is very much worth a visit (Photos 3-22, 23). Good columnar jointing is seen at 3 o'clock.

② To make a detour to the dry waterfall, prepare to turn off to the east on Cinder Road, just over 3 miles north of Little Lake. The junction is marked by an intersection sign, and the turnoff is paved. Follow this road east for 0.5 mile, take the

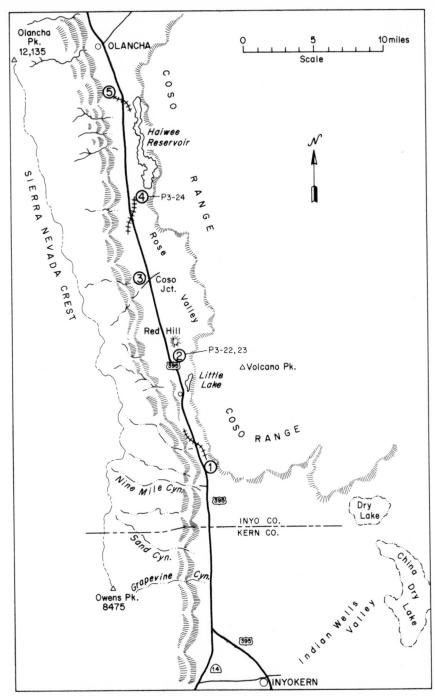

Figure 3-14. Segment M, U. S. 395 junction to Olancha.

Photo 3-20. U. S. Highway 395 as inundated by a debris flow just south of Little Lake in 1956. (Photo by Pierre St. Amand).

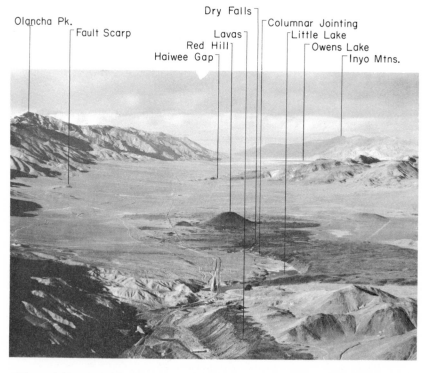

Olancha Pk.   Dry Falls ⌐
⌐Fault Scarp   ⌐Columnar Jointing
Lavas ⌐   ⌐Little Lake
Red Hill⌐   ⌐Owens Lake
Haiwee Gap⌐   ⌐Inyo Mtns.

Photo 3-21. Looking north over Little Lake. (Air photo by Roland von Huene).

132

Photo 3-22. Looking upstream to lip of upper dry fall of glacial Owens River, north of Little Lake. Figure at lip gives scale.

Photo 3-23. Looking down stream-cut gorge below upper dry fall near Little Lake. Picnic party near lip of lower fall. Foreground rock surfaces scoured, polished, and potholed by rushing torrent.

133

first well traveled right branch which heads south, follow it 0.3 mile to a Not a Through Road sign and turn east on the side road there, and follow it 0.5 mile to a wide parking area in the lavas.

The orange dab on rocks at the east edge of the parking area marks the start of a trail which continues east across the uneven lava surface for a few hundred yards, a 10-minute walk at most. The course is indicated by further orange markings. Note that the lavas are full of little holes, ¼-½ inch across. These were made by bubbles of gas in the molten material.

The falls are 150 feet south of your encounter with the old stream course. They occurred in two drops totaling more than 100 feet, the second being about 200 feet farther south. Note that the rock surfaces have been scoured, smoothed, and fluted by the fast flowing, presumably silty waters (Photo 3-23). The pot holes were cut into the lavas by fixed vortices in the stream. Water flowed past this locality at several different times, but the features we see now were formed during the last major discharge, probably 10,000-15,000 years ago.

This was an attractive place in those days. Indians found it so, for they spent much time in the area, as shown by living sites and many artifacts uncovered by the excavations of Southwest Museum personnel. You can still find small chips of shiny obsidian (volcanic glass) scattered along the banks of the river channel. The obsidian comes from exposures in the Coso Range only a few miles to the northeast.

● As you return to the highway something is seen of the cinder mining operation on the south side of Red Hill. Volcanic cinders are not the remains of burned rock but rather small pebble-sized fragments of highly porous lava, usually red or black, explosively thrown out of a volcanic vent. Cones are formed as the cinders pile up around the vent. Red Hill has been extensively mined for lightweight aggregate used in concrete and cement blocks.

● Back on the highway and headed north, we traverse lavas for another 2 miles. After passing Red Hill, nondrivers get a good view east, at 4-5 o'clock into the Coso Range, of other volcanic cones and recent lava flows coming down canyons. South-bound travelers get even better views of these features.

● North of Red Hill the highway enters Rose Valley in the middle of which is a grove of cottonwood trees marking the location of Coso Junction (Oasis). Some ponding of glacial waters probably occurred in this valley, but this shallow body did not leave strong shoreline markings that we can see. The white area a mile or two west of Coso Junction marks a site where pumice (a frothy volcanic glass) from the Coso Range was formerly milled by the Coso Pumice Company.

③ Farther west, breaking the alluvial apron near the base of the Sierras, you can see a linear embankment, green in summer with grass and scattered trees, locally known as Portuguese Bench. This is a remnant of a fault scarp created by displacement along a fracture within the Sierra Nevada frontal fault zone. Geologically, this scarp is not very old, but chronologically it is old enough to have been largely destroyed by erosion and deposition in areas opposite the mouths of major canyons. Its remnants are preserved mostly in locations between canyon mouths.

④ Beyond Coso Junction at about 1 o'clock, another stream-cut gap becomes visible at the base of the hills. It transects a spur projecting west from the Coso Range and is the site of the dam and powerhouse of Haiwee Reservoir. It too was cut by the overflowing glacial Owens River. Keep an eye on this locality as we move north.

• Shortly we climb up a gentle grade out of Rose Valley, cross a railroad track (occasional trains, so be wary) and straighten out at a higher level. Now off to the east at 2:30 o'clock at the base of the Coso Range the features of a large rock slide can be seen just upstream from the mouth of the stream-cut narrows (Photo 3-24). The slide involved mostly brownish rock. Its head is marked by a high cliff, and the body of the slide is an irregular jumble of huge rock blocks. It was probably caused by over-steepening of the hill face through undercutting by glacial Owens River.

• At the rest stop, about 2 miles beyond the railroad crossing, one gets a good view east to Haiwee Reservoir (Photo 3-25) of the Los Angeles Aqueduct system which occupies a former meadow along the course of glacial Owens River. The white rocks in the hills behind the reservoir are part of the Coso Mountains Formation, a late Pliocene (2-3 m.y.) land-laid sedimentary deposit, which contains fossil remains of such animals as large horses, slender camels, hyena-like dogs, short-jawed mastodons, meadow mice, and rabbits. The region must have been a grassland at that time.

• Within a few miles note that the fan surfaces along the highway are bouldery, indicating a different kind of rock in the

Photo 3-24. Rock slide at narrows below Haiwee Reservoir dam. Cliff is break-away scarp at head; irregular, road marked mass below is body of slide.

Old Shoreline
Little Lake
San Andreas F.
Garlock
Fault
Haiwee Reservoir
Owens Lake
Olancha
Sierra Upland
Olancha Peak

Photo 3-25. High-altitude oblique view south-southwest over Owens Lake. (U. S. Air Force photo taken for U. S. Geological Survey, 374R-184).

Sierras than near Mojave or west of Indian Wells Valley. Fault-scarp remnants are again seen on the west closer to the road about 5 miles from the rest stop.

(5) Shortly we recross the railroad, and in less than a mile, the open cement trough of the Los Angeles Aqueduct passes under the highway. Here, just a few hundred yards west is an old, ragged, bouldery fault scarp breaking the fan surface. The railroad traverses its face.

● Ahead are the usually dry flats of Owens Lake. Before Owens River was diverted into the aqueduct in 1913, this was a large, blue, salt lake of considerable beauty because of its setting in desert country at the base of towering mountains. In wet years water still covers part of the lake floor and may linger for a year or two before evaporating. The pre-1913 lake was about 30 feet deep and covered 100 square miles. Glacial Owens Lake which fed the overflowing glacial Owens River was closer to 220 feet deep and about twice as large.

● Coming into Olancha, the small dunes on the valley floor at about 2 o'clock have been built by south blowing winds, largely since Owens Lake dried up. Note how high and abrupt the Sierra front has become west of Olancha where Olancha Peak reaches an elevation of 12,123 feet; Olancha itself is at 3649 feet.

Segment N—Olancha to Independence, 38 miles, Figure 3-15

● About a mile north of Olancha the highway curves into a more northerly course, and the very bouldery surface of the Olancha Creek fan is seen ahead and to the west. The boulders, some the size of a small house, reflect the massive sparsely jointed nature of the rock source in the mountains and demonstrate the transporting power of debris flows. The abruptness and ruggedness of the Sierra Nevada front here is a reflection of the massiveness of the granitic bedrock.

● Owens Lake was saline, and in times past chemical plants have recovered salts from brines and from evaporated residues on the lake flat. The principal compound produced was sodium carbonate or soda ash. Cartago, ahead, was once the location of such operations, and white waste piles there mark the sites of former chemical plants. (Note odometer at Cartago.)

● At the railroad crossing 2.5 miles north of Cartago, the broad low saddle on the eastern skyline separates the Coso Mountains from the Inyo Range on the north. At 2:30 o'clock is Malpais Mesa, an area of nearly flat dark lava flows dissected by canyons and dropped down in a series of step faults. The lavas are but a thin veneer on older rocks.

● The front of the Sierra Nevada begins to change character opposite Cartago, and by the time we get to Cottonwood Creek, 6.5 miles from Cartago, the range front is less abrupt, less rugged, and has a local mantle of loose weathered debris. This mantle can be seen in the sharp gullies on the mountain face back of Cottonwood Creek powerhouse, about 2 miles north of Cottonwood Creek. These were cut when a flume high on the hillside broke. This range-front difference is caused by a change in bedrock, from the massive, clean, Sierra granite back of Olancha, to an older, messed-up mixture of granitic and metamorphic rocks that weathers more easily, from Cartago north.

● After passing Cottonwood Creek powerhouse road, look to the eastern skyline in the Inyos at about 2:30 o'clock. That is the approximate location of the famed Cerro Gordo mine, one of California's richest silver-lead producers. It enjoyed a heyday in the 1870's and produced bullion valued at about 17 million dollars—1870 dollars not 1970 dollars.

● In another 2 miles, 11.2 miles from Cartago, is an abandoned chemical plant just east of the highway at Bartlett. It used brines pumped from wells far out on the lake floor and evaporated in the rectangular, diked areas south of the plant. Soda ash and boron was recovered from these brines. From here, on a clear day, you may be able to see the buildings at Keeler far across on the east shore at 3 o'clock. Keeler has been the longest enduring of all chemical operations on Owens Lake, salts being first harvested there in 1885.

① About 3 miles beyond the chemical plant on the near floor of the valley at 1-2 o'clock are a number of abandoned parallel shoreline marks of historic Owens Lake (Photo 3-26). Like rings on a bath tub these formed as the water fell following diversion of Owens River in 1913. Higher shoreline features were formed 10,000-15,000 years earlier when the lake was more than 200 feet deep and overflowed to the south. Remnants of this old shoreline are seen best along the east side, south of Keeler (Photo 3-25). They are hard to recognize in the area just traversed because

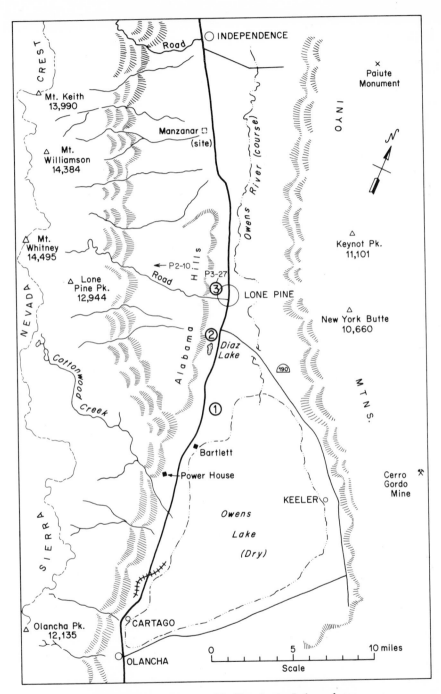

Figure 3-15. Segment N, Olancha to Independence.

of erosion and deposition by the powerful fan building streams coming out of the Sierras.

● In another mile the east facing scarp of the Alabama Hills begins to rise just west of the road (Photo 3-26). Where first seen it consists of boulderly fan gravels, but these soon give way northward to bedrock.

● As we motor north, look east across the valley to the west face of Inyo Mountains. They are composed largely of rocks unlike those making up the Sierra face. They are darker, and more variegated, and in places you should be able to make out steeply inclined bedding (layering). These layered rocks are sedimentary deposits of Paleozoic age, those seen ranging roughly from 250

to 500 m.y. old. Also present are some Mesozoic metavolcanic rocks, perhaps 150-200 m.y. old, and some still younger Mesozoic igneous intrusive rocks resembling those of the Sierras.

● Keep an eye on the Alabama Hills. Their scarp increases in height and shortly becomes a bedrock face, composed first of dark Triassic metavolcanic rocks (200 m.y.) which give way to more massive rugged outcrops of Sierra granitic rocks back of Diaz Lake, a pond west of the highway approaching Lone Pine (Photo 3-26).

② Diaz Lake occupies a sunken strip between two fault scarps formed during the Owens Valley earthquake of 1872. One forms a low, east-facing, linear bank about

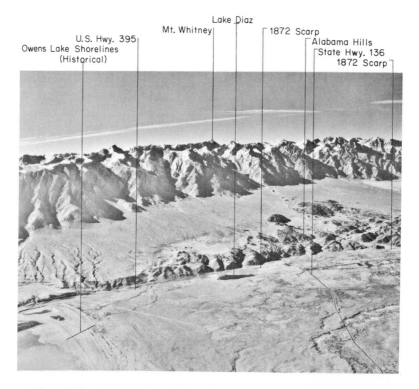

Photo 3-26. Low-altitude oblique view of Alabama Hills and east face of Sierra Nevada near Lone Pine. (U. S. Geological Survey air photo, GS-OAI-5-17).

10-15 feet high extending along the west side of the north end of Diaz Lake. It extends into the meadow to the north where trees grow on its face. The clubhouse at Lone Pine Golf Course sits above the scarp. Traces of the other scarp, which faces west, are harder to see from the highway.

● Mt. Whitney is well viewed on the distant skyline over the Lone Pine Golf Course, or better still, from opposite the Lone Pine Airport, where you can pull off the highway. Don't confuse Lone Pine Peak (Photo 2-10), just under 13,000 feet, which looms so large on the left because of its range front position, with sharper and farther back Mt. Whitney (14,494).

● Upon entering Lone Pine, it is time to consider the nature of Owens Valley. It is a keystone shaped block dropped down between two huge fault zones, one near the base of the Sierras and one along the Inyo Mountains front. Such a block is termed a *graben* by geologists.

However, this is clearly an oversimplified picture, for we have just traveled several miles along the face of a fault scarp well out on the floor of the valley. The east face of the Alabama Hills is but the top of a bedrock scarp nearly 10,000 feet high which is largely buried by over 9000 feet of alluvium and other unconsolidated deposits. This means that the bedrock floor of Owens Valley east of Lone Pine is 6000 feet below sea level, ample evidence that the Owens Valley block has truly been dropped down and not just left behind as the Sierras and Inyos were elevated. All this movement has occurred within the last few million years, and, in fact, still goes on.

● Years ago some traveling medicine man misled the people of Lone Pine into advertising the rocks of the Alabama Hills as the oldest in the world. This is pure nonsense. The oldest rocks in the hills are Triassic metamorphics, only about 200 m.y. old. Just across the valley in the Inyos are rocks 500 m.y. old; farther north in the White Mountains, rocks approach a billion years in age, and east in Death Valley, rocks as old as 1.8 billion years have been measured. Lone Pine has a rich spectrum of attractions; it doesn't need old rocks. Just travel some of the back roads in the Alabama Hills to find that out.

● The 1872 earthquake was one of the largest of historical time in California, possibly slightly greater than the San Francisco shock of 1906. Considering the sparse habitation in the valley at the time, the twenty-three to twenty-nine deaths reported represented a significant percentage of the population. Many fault scarps were formed in 1872, and we have begun to see them in the environs of Lone Pine. They continue as far north as Big Pine. The main line of breaking was along the Alabama Hills fault and its northward continuation, not on the Sierra frontal fault.

③ As we leave Lone Pine, record ◄ your odometer reading opposite the County park displaying the jet fighter plane at the north edge of town. About 0.2 mile north of the park at 8 o'clock you should catch a glimpse of one of the more spectacular scarps formed during the 1872 quake. It is a linear, light gray, bouldery face 20 feet high cutting the fan surface just this side of the base of the Alabama Hills (Photo 3-27).

● A little farther north we pass the grave site of some 1872 earthquake victims. The graves are west of the highway on top of a steep rise which is an older fault scarp. The 1872 scarp is well seen ¾ mile southwest from the grave site. The highway runs along the base of the older scarp for about 1.5 miles, and other segments of old scarp are seen farther west of the highway for another mile north. Beyond that point closer to the

Photo 3-27. Fault scarp of 1872 just west of Lone Pine.

hills you can see the fresher scarps, formed in 1872, cutting across the surfaces of small alluvial fans. Don't mistake the embankment of the Los Angeles Aqueduct for a fault scarp.

● You will note a meadowy area dotted with tree stumps and old dead tree snags just east of the road about 3.5-4 miles north of Lone Pine. This area was flooded during the 1872 'quake when Owens River was diverted into this spot creating a shallow pond.

● Nearly 4.5 miles from Lone Pine park the highway climbs up and over the 1872 scarp, and in less than a mile passes the spillway of Los Angeles Aqueduct.

The front of Alabama Hills north from Lone Pine is again made up of the darker Triassic metavolcanic rocks that yield smoother slopes than the more massive Sierra granitics seen back of Diaz Lake.

● As we emerge from behind the Alabama Hills, the full majesty of the Sierra front appears. The crest of the range towers 10,000 feet above us and 8000 feet above the heads of the alluvial fans. From here to the Poverty Hills a little south of Big Pine—this is a spectacular example of a major fault scarp.

● West at 9 o'clock from the abandoned site of Manzanar (dead apple orchards and other trees, 10 miles north of Lone Pine) is Mt. Williamson (14,375 ft.). Southbound travelers get a superb view of this huge mountain all the way south from Poverty Hills.

● Approaching Independence, note that the crest of the Inyo Mountains to the east has become somewhat lower and less rugged. This happens because the core of the Inyos is here made up of a large granitic intrusive body that weathers readily and uniformly. You can make out the approximate outline of this body by its lighter color and gentler terrain, compared to the darker rocks into which it has been intruded.

● The line of the 1872 fault break passes about 3 miles east of Independence and is marked by a prominent scarp in the alluvium. To see it, turn east on Mazourka Canyon road at the south edge of town, and then drive 3.1 miles

● You may be aware that the original course followed by Owens River is indicated by the line of trees far off along the east side of the valley. A good map would show you that the river course lies close to the base of the Inyo Mountains, not in the middle of Owens Valley. It has been shoved eastward by the huge fans built by streams flowing out of the Sierras. These streams are much better supplied with water and debris than the smaller intermittent streams flowing westward out of the Inyos.

Segment O— Independence to Bishop, 43 miles, Figure 3-16

▶ • (Note your odometer opposite the county courthouse in Independence.) Northbound out of town careful inspection of the eastern skyline should reveal Paiute Monument or Winnedumah. This is a natural tower about 80 feet high formed by erosion of jointed granitic rock.

• Note that the front of the Sierras is here marked by uncharacteristic projecting spurs and forelying rock knobs. We tend to think of the Sierra frontal fault as a single, simple fracture. Actually, it is a complex zone consisting of many fractures, mostly parallel. However, locally, as for example here, fault relationships are more complex, and the topography of the mountain face is more irregular. In truth, from here north the Sierras' face becomes increasingly less linear. Stretches of high straight scarp are offset from one another or separated by projecting rock masses.

• At the end of the divided highway, roughly 4 miles north of Independence, we see ahead on both sides of the valley (at 10-11:30 o'clock and at 2 o'clock) the cinder cones and lava flows of the Taboose-Big Pine volcanic field (Photo 3-28). Red Mountain at 11:30 o'clock is one of the larger volcanic cones. These features cannot be very old, as some of the lava flows rest on alluvial fans, and even the cinder cones are not much modified by erosion. It is thought that the Sierra and Inyo frontal faults extend deeply into the earth's crust providing a means of access to the surface for molten lavas. The composition of the lavas, and of some foreign inclusion within them, suggest that they came from great depth. Lava and cinders were spewed from perhaps thirty or more individual vents, many of which are aligned along recognizable faults. Eruptions occurred in at least four separate episodes all within perhaps the last 100,000 to 200,000 years.

① About 8 miles north of Independence, 4 miles north of the end of the divided highway, keep a careful watch on the base of the Sierra scarp. At 9 o'clock is a narrow V-shaped slot cut into the steep mountain face by Sawmill Creek. If the light is right you will see masses of black rock adhering to both walls of this slot a little above its mouth. These are remnants of lava that flowed down the canyon filling it to a depth of 150 feet. Stream erosion has since cut a narrow cleft through the lava. These flows are interesting to geologists because they have been dated as roughly 100,000 years old by the potassium-argon method. Farther up the canyon a glacial moraine rests on top of the lava, showing that this stage of glaciation occurred less than 100,000 years ago. Still farther up Sawmill Creek the same lava unit rests on top of another morainal deposit which, accordingly, must be older than 100,000 years.

• Shortly the highway passes onto some lava flows, and we get a closer look at their irregular surface. If you stop, you will see that the lavas are twisted, jagged, blocky, and locally full of small holes formed by bubbles of gas trapped in the solidifying rock. These are what geologists call basaltic lavas, relatively low in silica and high in iron and magnesium.

• North of the rest area, about 11 miles from Independence, Poverty Hills lie ahead at 11 o'clock. They are composed of older crystalline rocks including granite and marble.

• Look east to 2:45 o'clock in passing the Aberdeen Station-Taboose Creek Campground road (14.5 miles from Independence) to see a red cinder cone near the base of the Inyos surrounded by an apron of

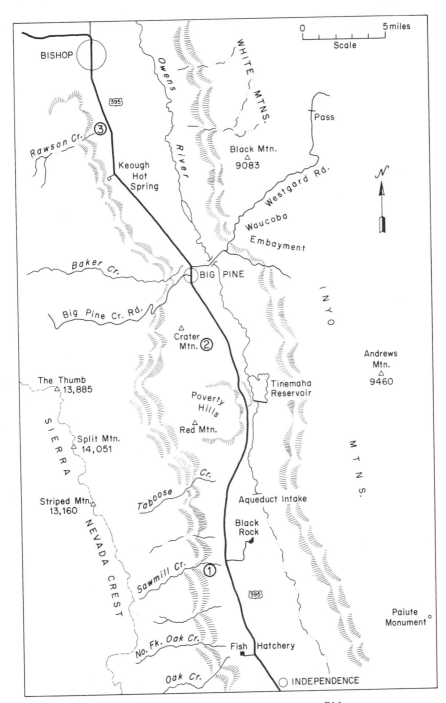

Figure 3-16. Segment O, Independence to Bishop.

darker lavas. Cinders are often red, their iron being oxidized while still hot during flight through the air or by gases associated with the eruption.

• About a mile north of the end of the four-lane highway across Poverty Hills, look west at 9 o'clock for a good view of a fresh red cinder cone on the far side of the green (in summer) fields.

② In another mile at about 10:30 o'clock, the large mass of nearby Crater Mountain (Photo 3-28) begins to claim attention; keep an eye on it. Crater Mountain is actually a granitic knob with lavas draped over its summit and flanks like chocolate on a sundae. Farther north you will see two knobs of light-colored granitic rock peeking through the lavas along its north ridge. If the sun is in the west, you may be able to see fault scarps extending northward across the lower eastern slope of Crater Mountain from 10 to 11 o'clock. The most continuous scarp was formed during the 1872 earthquake (Photo 3-28).

• Approaching Big Pine you get some excellent views in clear weather and late afternoon lighting of the variegated sedimentary rocks composing the Inyo Mountains front to the east. Viewing is particularly good for southbound travelers. Most of the sedimentary rocks exposed there are of Cambrian age, 500-600 m.y. old.

• Big Pine Canyon, west of town, is hard to pick out because its mouth is plugged by large massive glacial moraines formed by the Big Pine glacier. If you drive up Big Pine Creek, the road switches back on the inner face of a moraine just after the second creek crossing, and your subsequent route nearly all the way to Glacier Lodge is upon lateral-moraine ridges. Glaciers reached the east base of the Sierras at many canyons north from here but at only a few to the

south, the southernmost recognized being Independence Creek.

• (Record odometer reading in the center of Big Pine.) About 0.5 mile north of town, the road to Westgard Pass, Deep Springs Valley, and the northern end of Death Valley takes off east. The subdued Inyo Mountains front and the lower skyline on the east are an expression of the Waucoba embayment which separates the Inyo Mountains from the White Mountains on the north (Photo 3-29). The white patches on the lower slopes of the embayment are uplifted lake beds not more than 1-2 million years old. Similar materials probably underlie the alluvium of the Owens Valley floor, as the unconsolidated fill in the valley here approaches a thickness of 8000 feet. The deepest part of the bedrock valley is close to the base of the Inyo and White mountains, indicating eastward tilting of Owens Valley.

• At Reynolds Road, nearly 2 miles out of Big Pine, look to the crest of the Sierra Nevadas at 9 o'clock. The ragged skyline ridge is part of the Palisades. The Palisade Glacier (Photo 3-29) lies out of view to the left at the foot of this cliff. It is the largest and one of the more southerly glaciers in the Sierras.

The present small Sierra glaciers may not be shrunken remnants of ice age masses. Instead, they were probably formed within the last 3000-4000 years, for the following reasons. Many lines of evidence show that several thousand years ago the climate was warmer and drier than now. Owens Lake may have dried up completely then, suggesting that the Sierras were probably denuded of perennial ice and snow at that time. About 3000-4000 years ago the climate became wetter and cooler, and more snow accumulated in the Sierras forming

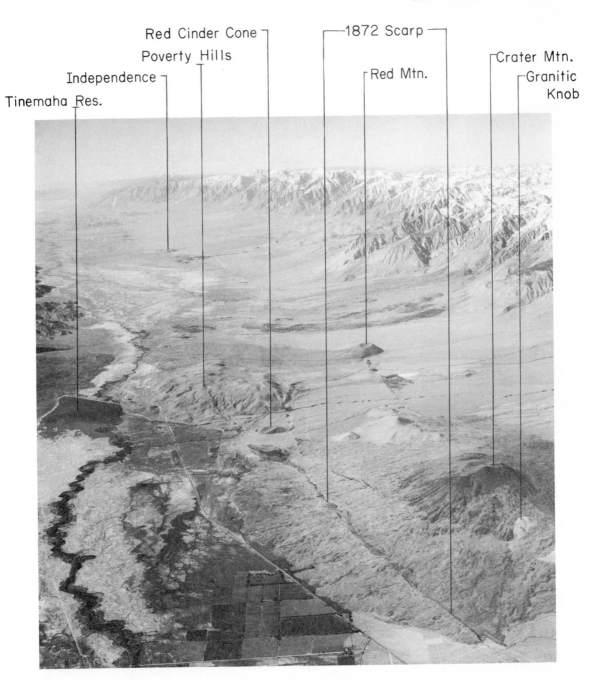

Photo 3-28. Looking south down Owens Valley from over Big Pine. (Air photo by Roland von Huene).

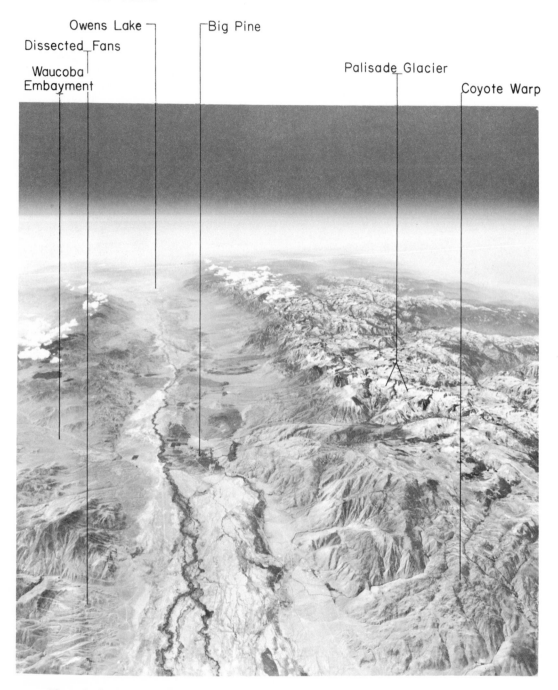

Photo 3-29. High-altitude oblique view south down Owens Valley from over Bishop (U. S. Air Force photo taken for U. S. Geological Survey, 018L-057).

perennial snow banks in protected spots. Eventually these snow banks grew into new glaciers. Judging from the moraines they have built, at least some of these glaciers attained their maximum length, about ½ mile, roughly between 1750 and 1890 A.D. Since then they have been shrinking.

● Out on the valley floor at 2 o'clock the "ears" of the Caltech Radio Astronomy facility are visible. They consist of three huge dishes (parabolic antennae), two 90 feet and one 150 feet in diameter, mounted on rails to facilitate shifting for a sort of "stereo" listening.

● About 8 miles north of Big Pine, good views are seen of the White Mountains to the east, especially in afternoon light by southbound travelers. Along the base of the range are some deeply dissected alluvial fan deposits extending far up some canyons. These uplifted fan deposits are another indication of the recency of geological deformation here. Keep watching them as we move north toward Bishop, looking especially to 3:30-4 o'clock north of the outdoor theatre.

Most of the rocks seen in the west face of the White Mountains from here to Bishop are Cambrian (500-600 m.y.) sedimentary beds. The east side and crest of the range farther north are composed of older, presumably late Precambrian sedimentary formations. Locally, mostly to the east and north, are much younger (100-150 m.y.) granitic intrusive bodies.

● One-half mile north of Keough's Hotspring, where the four-lane highway starts and Collins Road turns off to the west, the fan surface just west of the highway at 10-11 o'clock has a peculiar hilly appearance. You are looking at the back side of a little fault block with a scarp facing back (west) towards the Sierras. You would normally expect scarps associated with uplifted mountain masses to face toward the valley. These anomalous mountain facing scarps are probably created by minor settling adjustments in the valley block.

③ If you'd like to escape the rush of traffic and see a little more of this fault structure, turn west on Collins Road, follow it 0.7 mile west to Gerkin Road, and then turn north. Now plainly visible just to the east behind the powerline towers is the scarp. It dies out northward in less than ½ mile. Follow Gerkin Road to rejoin U. S. 395 a short distance ahead. Southbound travelers can make this loop by turning off on Gerkin Road just short of the outdoor theater and returning to U. S. 395 on Collins Road.

● The front of the Sierras from Big Pine north is less abrupt and imposing than farther south (Photo 3-29). This part of the Sierra face is regarded as having been bent (warped) more than faulted. As you will see shortly, it constitutes a sort of spur behind which the main mass of the Sierras is offset westward about 10 miles. This structure, called the Coyote Warp, bends down under Owens Valley to the east and plunges north under Bishop.

Section P—Bishop to Mammoth, 42 miles, Figure 3-17

• U. S. Highway 395 turns west at the Y on the north edge of Bishop (note odometer mileage here). Continuing west we see the 10-mile offset of the Sierra front behind the north end of Coyote Warp.

• West of Coyote Warp is the Bishop Creek country. By looking to the mountain base at 11 o'clock just beyond the Y you can make out the lower Bishop Creek gorge. The somewhat unusual north flowing course of Bishop Creek within the mountains is presumably determined by uplift of the west flank of the Coyote Warp. Immediately west of lower Bishop Creek are subdued, brush covered, gray hills through which the new Bishop Creek highway takes its course. These hills are composed of bouldery deposits laid down by an early advance of Bishop Creek glacier. They remain in view for the next 4 miles.

• Roughly 2.5 miles west of the Y, about half way up the face of Coyote Warp at 9 o'clock, is a smooth bench sloping gently northward. It is capped by a lava flow resting on an erosion surface cut across older granitic rocks. If the light is good and your eyes are sharp, you may be able to make out some large, light-colored boulders (granitic) on top of the lavas about where the trees begin. These boulders were placed there by an early glacier flowing out of the higher country to the south.

• Between 3 and 4 miles from the Y, the large mountain on the skyline at 11:45 o'clock is Mount Tom (13,652), and the lower and nearer rocky terrain is Tungsten Hills, named for the many tungsten mines and prospects therein. The large irregular embankments high up in the canyons on Mt. Tom are rock glaciers. A rock glacier is an accumulation of large angular boulders that moves or has moved slowly down slope by a creeping mechanism.

• At ½ mile beyond Ed Powers Road we start a gentle descent interrupted part way down by a flat step (terrace). Round Valley, ahead, occupies a down dropped block lying between this west facing rise, produced by a fault and a warp paralleling our route for the next 5 miles, and the east-facing Sierra-front scarp.

• In less than a mile Pleasant Valley Dam road turns off east down a wide valley which formerly carried a much larger drainage out of Round Valley. Within the next 2.5 miles the highway gradually converges to an intersection with this old stream course (Photo 3-30) at the point of rocks ahead. Approaching the point of rocks, the cliffs east of the old stream valley are composed of Bishop tuff, of which we see a great deal up Sherwin Grade ahead.

① Beyond the point of rocks and across the valley at 10 o'clock are the tips of glacial-moraine ridges protruding from the canyon of Pine Creek (Photo 3-31). Farther up Pine Creek is one of the largest tungsten processing mills in the United States. The mines supplying the mill are higher and farther back in steep, glaciated terrain near the headwaters of a north branch of Pine Creek at elevations ranging from 9500-12,000 feet. They cannot be seen from the highway.

The tungsten ore is in what geologists call "contact deposits." In this instance the contact is between old metamorphic rocks, including layers which were once limestone, and younger intrusive granitic rocks. Although the ore is in the highly altered metamorphic rocks, the principal ore elements, tungsten and molybdenum, were provided by hot fluids and vapors given off by the

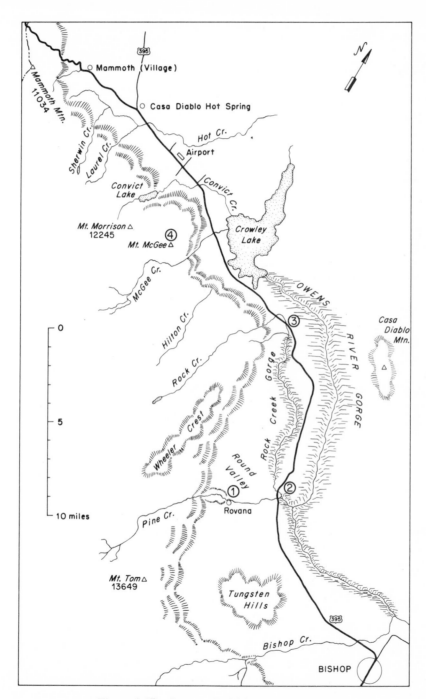

Figure 3-17. Segment P, Bishop to Mammoth.

intruding igneous body. This has long been one of the richest tungsten operations in the United States.

● The high, abrupt ridge of the Sierras west of Round Valley is Wheeler Crest. It is composed of massive granitic rocks with steep jointing that give strongly castellated cliffs. Keep an eye on this face, as we go north. (Read ahead in the guide.)

② Just before the highway swings slightly east to start the ascent of Sherwin Grade, a quick look east reveals the gorge of Birchim Canyon (Photo 3-30) through which Pine and Rock creeks flow out of Round Valley to Owens River. Earlier they probably followed the old stream valley seen 5 miles back. (Note odometer at the curve.)

● The ascent of Sherwin Grade takes place on the upper surface of the Bishop tuff. Somewhere not too far to the north, about 700,000 years ago, volcanic vents repeatedly ejected clouds of hot glowing ash and rock particles that spread out over the surrounding landscape for great distances. Heavy because of the suspended particles, these clouds flowed by gravity, and they were kept mobile by gases being released from the hot volcanic fragments. When the material came to rest, it was still so hot that the particles partly recrystallized and fused, creating a coherent rock. This "glowing cloud" was the same sort of phenomenon that wiped out the city of Saint Pierre when Mt. Pelée erupted on the island of Martinique, in the West Indies, in 1903.

When the Bishop tuff eruptions ceased, a nearly level sheet of material averaging 500 feet in thickness flooded and partly submerged an originally rough landscape leaving islands of older rock sticking through. The sheet extended at least 50 miles from Bishop to Mono Lake. Subsequently, it has been deformed by faulting (Photo 3-30)

and warping, and then it was eroded, so that only remnants are preserved today. We are traversing the largest remnant, and the gentle incline we ascend is the result of later tilting.

You can see what the Bishop tuff looks like as we pass through road cuts going up Sherwin Grade. In fresh exposures it is mostly pink, but it weathers brown, and the white zone near the top of some cuts was formed by weathering and deposition of calcium carbonate along fractures. Close inspection of the rock shows that it contains fragments of pumice, shiny particles of feldspar and quartz, a lot of dull, nondescript ashy material, and small chunks of a variety of older rocks, mostly volcanic.

● As the highway rises onto the surface of the tuff, a mile from the start of the grade, start looking left and right, especially right. You will see many small pimple-like hills, often in clusters (Photo 3-30). They represent places where residual gases in the deposited materials found their way to the surface hardening the tuff. Erosion has since converted these firmer spots into little conical hills, 20-75 feet high.

● About 2.3 miles up the grade the highway enters a long straight stretch, and ascending it we see Casa Diablo Mountain about 5 miles off at 1 o'clock. This particular "House of the Devil" is an island of older granitic rock rising through the Bishop tuff. On the far skyline at 2 o'clock is White Mountain Peak (14,246 ft.), composed of old metamorphosed volcanic rocks intruded by younger granitic bodies. A cluster of pimple hills is seen on the tuff surface in the mid-distance at 9-11 o'clock.

We are here traveling a course roughly parallel to and about midway between the gorges of Rock Creek, to the west, and Owens River to the east (Photos 3-30, 31). These streams have abruptly cut hundreds

Photo 3-30. High-altitude vertical view of Sherwin Grade area; north to right, scale in lower left. (U. S. Air Force photo taken for U. S. Geological Survey, 018V-058).

Pine Creek
Moraines

Wheeler Crest

McGee Ck. Moraine

Little Round Valley

Tom's Place

U.S. Highway 395

Sherwin Till

Owens R.

Gorge

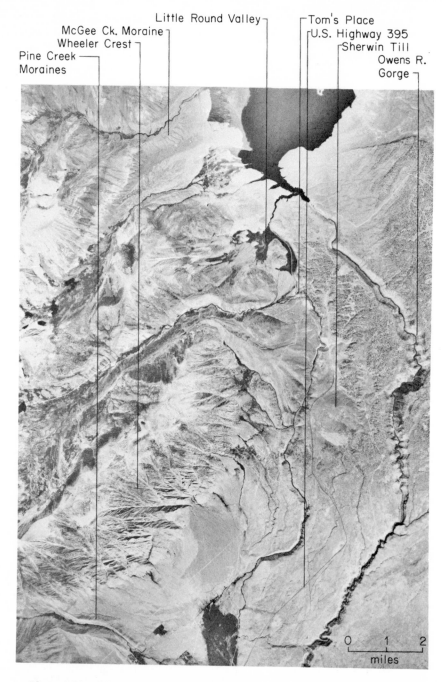

0   1   2
miles

Photo 3-31. High-altitude vertical view of upper Sherwin Grade-Crowley Lake area; north to upper right corner, scale at lower right (U. S. Air Force photo taken for U. S. Geological Survey 018V-059).

of feet into the Bishop tuff and in places into the underlying rocks. For those interested, spectacular views of the tuff can be seen in the walls of Owens River gorge, now largely dry owing to diversion of water through a series of powerhouses by the Los Angeles Bureau of Water and Power. Side roads taking off east near the top and bottom of Sherwin Grade lead to the gorge.

● The highway eventually curves west, and about where the piñon trees begin you get a good view at 8-9 o'clock toward Round Valley. You can now see more clearly the morainal ridges protruding from Pine Creek canyon.

Southbound travelers get a fine view of Mt. Tom as they come down Sherwin Grade, and also of Pine Creek and Round Valley. Note the rock glacier in the canyon high up on Mt. Tom.

● Just before the divided highway begins (8 miles up the grade) we pass out of the Bishop tuff onto some granitic rocks and then in less than a mile into cuts exposing coarse bouldery material. This is a noteworthy glacial deposit, appropriately named Sherwin till. Till is unsorted bouldery material deposited directly from glacial ice.

● At the Sherwin Summit sign (7000 ft.), you again see cliffs of the Bishop tuff at 1-3 o'clock. It overlies the Sherwin till which composes subdued brush covered slopes east of the highway and south of the cliffs.

● White material seen in the second road cut, 0.5 mile beyond the summit, is pumice which lies beneath the Bishop tuff.

③ A mile beyond the summit as we descend to Rock Creek and pass through a very deep road cut exposing significant geological relationships, watch the east side. The white material is pumice beneath the Bishop tuff. Toward the north end of the cut you can see that the pumice overlies a coarse bouldery deposit, which is Sherwin

till. Mineral crystals in fragment of the pumice from this cut have been dated by the potassium-argon method as slightly more than 700,000 years old. This fixes the age of the Sherwin glaciation at about 750,000 years old, for the till had been extensively weathered and eroded before the pumice was deposited. I always make my family observe a reverent moment of silence as we pass through this cut in recognition of its geological significance. Unfortunately, it will eventually be bypassed in a rerouting of the highway, but the cut is to be preserved as a point of geological interest.

● After crossing Rock Creek we shortly enter a new section of four-lane divided highway. Look quickly at 9:30 o'clock to see the rock gorge at the mouth of Rock Creek. The glacier that deposited the Sherwin till came from Rock Creek, but it did not come through that gorge, which is post-Sherwin in age. It followed a course along Whiskey Canyon about 1.5 miles south.

● Rock Creek now makes a sharp turn to the south just east of the Crowley Lake Drive exit (0.6 mile ahead) to flow into its lower gorge and down to Round Valley. At an earlier stage, it continued northeastward about parallel to the new four-lane highway through Little Round Valley and into Owens River close to Crowley Lake dam site.

● About a mile beyond the Lake Crowley Drive exit is a nice contrast between jointed, gray granitic rock in the large knob west of the highway and massive, tan Bishop tuff in cliffs east of the highway. The ditch just west of the road here is used at times of high water to divert part of Rock Creek along its old route into Crowley Lake so that the water can be run through the powerhouses and have all the kilowatts taken out. Within another mile we enter Little Round Valley. The narrows at its north corner, where the highway cuts up into the hills, and where you can see an arm of Crowley Lake, are along the old abandoned Rock Creek course.

● All of the road cuts through the hills beyond Little Round Valley are in Bishop tuff.

● Just beyond the underpass at the Crowley Lake-Hilton Creek exit, look west to the mountain front at 9 o'clock to see the massive Hilton Creek moraine, a huge embankment of bouldery material on the mountain front. This might be called a dump moraine. Hilton Creek glacier came to the lip of its hanging valley and just dumped its load of debris down over the steep mountain face below.

● Ahead at 10-11 o'clock on the far side of the brush-covered flat is the spectacular right-bank, lateral-moraine complex of McGee Creek, fully 700 feet high (Photo 3-32). Right and left banks of streams or canyons are designated as though facing downstream. This moraine was built during several stages of glacial advance.

Photo 3-32. Late afternoon view of fault scarp cutting glacial moraines at mouth of McGee Creek. (Air photo by Roland von Huene).

Note that the rocks of the Sierra Nevada front here display many shades of brown, red, white, and black. These variegated rocks are part of a large metamorphic septum. Originally they were shales, sandstones, and limestones, and some contain identifiable fossils showing that the oldest layers were deposited about 400 m.y. ago.

● Crowley Lake, to the east, is artificial, but about 75,000 years ago a large natural lake about 200 feet deep lay in this basin covering an area of 90 square miles. The outflow from this lake did much of the cutting of Owens River gorge.

Crowley Lake occupies the south part of Long Valley which is the first of several structurally depressed basins along the east front of this sector of the Sierras. Mono Basin and Bridgeport Basin are two more lying farther north. These are geologically recent structures deeply filled with unconsolidated deposits. The fault cutting the moraines at McGee Creek (Photo 3-32) is one expression of the geologically recent date of some of the deformation in Long Valley.

● At McGee Creek look to the eastern skyline at 2:45 o'clock to see Glass Mountain, so named because the volcanic rock composing it is so rich in glass (obsidian).

● After rounding the corner and crossing Convict Creek, about a mile to the southwest is the massive morainal system of the Convict Creek glacier. Study Photo 3-33 to get a picture of relationships here. The road to Convict Lake (1 mile ahead) takes off southwest and crosses a large lobate morainal mass formed by the latest glacial advance. At an earlier stage the Convict glacier followed a more southerly route and built along its left side the huge lateral moraine, nearly 1000 feet high, which you see by looking back at 8 o'clock from the Convict Lake turnoff. Southbound travelers

get a better and easier view of these Convict moraines.

Along here we get a good look at lone-standing Mammoth Mountain at 11 o'clock on the skyline. If it's winter and everything is normal, the mountain will probably be crowned by its own local cloud.

Incidentally, this is "True Grit" country. Those of you who are up on your geology should have recognized outcrops of the Bishop tuff in parts of that movie, as well as scenes along Hot Creek, a little north of us. You should also have spotted Mt. Morrison from time to time in the background. It is the high pointed skyline peak at 8:45 o'clock, made up of steeply inclined layers of metamorphic rock. However, much of the film was taken in Colorado; so don't search around here for those gorgeous groves of quaking aspen.

④ Beyond the Convict Lake turnoff, a look back to about 8 o'clock on the skyline (2 o'clock for southbound travelers) may show you a deposit of large, light-colored boulders resting on darker rocks about where the trees are thickest, provided everything isn't covered by snow. This is the McGee till (Photo 3-33), one of the oldest glacial deposits of the Sierras. The boulders are granitic, some up to 20 feet across, which were carried from upper McGee Creek and deposited on the crest of this ridge by the McGee Creek glacier long, long before the moraines at the mouth of the present canyon were deposited.

● At the State Fish Hatchery Road, the white spot on the far hills at 3 o'clock is dust from a clay plant that gets its material from altered volcanic rocks in the hills to the north.

● About 0.5 mile beyond the Fish Hatchery turnoff, knobs of dark lava start appearing close to the road on both sides. We are riding up onto the back of some partly

Old Lateral Moraine
McGee Till

Mt. Morrison

Volcanic Mesa

Convict Lake

Hot Cr. Rd.

Young
Lobate
Moraine

Photo 3-33. Looking south to Convict Lake over Convict Creek glacial moraines. (Air photo by Roland von Huene).

buried lava flows coming out of the Mammoth embayment into which we turn shortly. The ridges and hills north of the highway here are also volcanic, but older.

● In another mile the highway curves and we then see clearly at 9:30 o'clock the steep "dump" moraine of Laurel Creek, up which a road switches back and forth.

● Approaching the Mammoth cloverleaf and exit, Casa Diablo Hot Springs lies ½ mile north at 1 o'clock. This is a geological hot spot with natural steam vents where some 10 test wells have been drilled to assay the steam resources for power generation, as now practiced at the "Geysers" north of San Francisco. Sometimes some of these wells leak; so you may see a plume or two of steam, particularly in winter.

● After turning off U. S. 395 we head west into the Mammoth embayment, and within a few hundred yards cliffs of black lava outcrop close by and continue on both sides of the highway for another mile. Different sequences of flows in these lavas have been dated as 200,000 and 300,000 years old. They rest upon bouldery glacial deposits judged to be 400,000-500,000 years old, a product of the Casa Diablo glaciation.

● A little over a mile from the cloverleaf we ride up on top of the flows, and shortly you will see large granitic boulders resting on the lava. These were deposited by a much younger phase of glaciation. The heavily timbered slope on the far side of the valley at 8-9 o'clock is the front of a massive bouldery moraine pushed out into the embayment by a tributary glacier from Sherwin Creek. Eastbound travelers see it and the steep brush-covered Laurel Creek dump moraine beyond more clearly.

● For those of you who drive on up to Mammoth Mountain from Mammoth Village, the rocks north of the road are lavas composing volcanic domes. You will also see a lot of fragmental pumice in road cuts, if they aren't covered by snow. This region has repeatedly been blanketed by falls of pumice thrown out of explosive vents in the Inyo and Mono craters to the north. Some of the pumice on Mammoth Mountain is about 1500 years old, as determined by a radio-carbon date on a buried tree stump. Explosion pits in the Inyo Craters, just a few miles north, are dated at only 650 years ago by the same means.

● The "earthquake fault" about 2 miles up the Mammoth Mountain-Devil's Postpile road is a good open crevice in a lava flow but not a very good example of a fault. A fault is a tightly closed fracture along which the crustal blocks on the opposing sides have slipped past each other. There are some suggestions of slip movement in the form of scratches on the walls of this crevice, but most of the movement has been a splitting apart, which is abnormal for a fault. Whether the splitting caused an earthquake is purely conjectural since the event was prehistoric.

● Mammoth Mountain is a complex volcanic dome. Some of its lavas are as much as 400,000 years old, but others are younger. If you ski the north face of the hill at chairlift 3 and work far enough east, you may be able to spot a steam vent or two about half way down. Mammoth Mountain is still warm and is best regarded as a dormant, not necessarily dead, volcano.

● Views from the high chairs and gondola at Mammoth Mountain are superb, especially west and north into the Middle Fork of the San Joaquin River, to the Minarets and the Ritter-Banner country. The low ridge extending north about 1.5 miles west of Mammoth Lodge is the drainage divide of the Sierra Nevada. About 3 miles north along its crest is an exposure of the oldest glacial deposits known in the Sierra Nevada.

These coarse bouldery gravels are sandwiched between two series of lavas which are dated respectively as a little more than and a little less than 3 million years old. The lava underlying the till composes the dark reddish brown cliff on the east side of the ridge, visible over the roof of Mammoth Lodge. The overlying lava is the one making up the gray outcrops on the peak rising to the skyline above the dark cliff.

The Mammoth region is rich in other geological phenomena and relationships. If you spend any time here and have an interest in natural features and history, get Genny Schumacher's excellent guide on the Mammoth Lakes. It's authentic, reliable, and well written.

APPENDIX A
GEOLOGICAL TIME SCALE

Era	Period	Epoch	Tentative Absolute Age
Cenozoic	Quaternary	Holocene	11,000 yrs.
		Pleistocene	2 million yrs.
	Tertiary	Pliocene	12
		Miocene	26
		Oligocene	37
		Eocene	53
		Paleocene	70 m.yrs.
Mesozoic	Cretaceous		135
	Jurassic		190
	Triassic		230 m.yrs.
Paleozoic	Permian		280
	Pennsylvanian		
	Mississippian		350
	Devonian		400
	Silurian		430
	Ordovician		500
	Cambrian		600 m. yrs.
Precambrian			600-3600 m.yrs.
---------------------------------- Lost Interval ----------------------------			
Origin of Earth			4600 m.yrs.

APPENDIX B
GLOSSARY

ABRASION The mechanical wearing of solid materials by impact and friction.

AGGLOMERATE A fragmental volcanic rock consisting of large, somewhat rounded stones in a finer matrix, much like conglomerate in appearance but wholly volcanic in constitution.

ALLUVIUM Unconsolidated gravel, sand, and finer rock debris deposited principally by running water; adjective *alluvial.*

ANGULAR UNCONFORMITY An arrangement in which older deformed stratified rocks have been truncated by erosion and younger layers have been laid down upon them with a different angle of inclination.

ANTECEDENT STREAM One which maintained its course in spite of localized uplift across its path; the stream anteceded the structure.

ANTICLINE A fold in stratified rock convex upward. Beds on the flanks are inclined outward.

ANTICLINAL CORE The mass of older rock in the heart of an anticline.

ANTICLINAL NOSE The place where beds at the axis of a plunging anticline pass beneath the ground surface.

ARROYO The wide, flat-floored channel of an intermittent stream in dry country.

ASH See volcanic ash.

AXIS The central line of an elongated geological structure such as an anticline or syncline.

BARRANCA A vertical walled gully cut by an intermittent stream in relatively unconsolidated material.

BARCHAN An isolated, crescent-shaped dune, convex upwind.

BASALT A fine-grained black lava relatively rich in calcium, iron, and magnesium. The extrusive equivalent (in composition) of gabbro.

BASEMENT Old crystalline rocks upon which younger rocks have been deposited.

BATHOLITH A very large igneous body intruded into the earth's crust at considerable depth where it cooled slowly to form coarsely crystalline rock.

BEDDING The layered structure of sedimentary rocks.

BEDROCK Consolidated rock material of any sort.

BENCH A level or gently sloping area interrupting an otherwise steep slope.

BENCH MARK An established mark, the elevation of which is accurately determined with respect to sea level.

BRECCIA A rock containing abundant angular fragments of rocks or minerals. These are sedimentary breccias, volcanic breccias, tectonic breccias, landslide breccias, and other types.

CARBONATE ROCKS Those composed of the minerals calcite (calcium carbonate) and dolomite (calcium-magnesium carbonate).

CALCAREOUS Rich in calcite.

CALCITE A common mineral composed of calcium, carbon, and oxygen ($CaCO_3$). The principal constituent of cement.

CALICHE A calcareous deposit formed within dry-region soils by weathering.

CAPTURE See stream capture.

163

CLEAVAGE   The facility to break along parallel smooth planes, especially in minerals, but also in rocks.

CONGLOMERATE   A sedimentary rock consisting of larger rounded rock and mineral fragments embedded in a finer, usually sandy matrix and all cemented together.

CRYSTAL   A regular, solid, geometrical form bounded by plane surfaces expressing an internal ordered arrangement of atoms.

CRYSTALLINE   Substances having fixed internal atomic arrangements.

CRYSTALLINE ROCKS   A term commonly applied to mixed igneous and metamorphic rocks, or to either separately.

DEBRIS   Broken-up and usually partly decomposed rock materials.

DEBRIS CONE   A cone-shaped accumulation of rock debris at the mouth of a gully or small canyon, usually smaller, steeper, and often rougher than an alluvial fan.

DEBRIS FLOW   A flow of usually wet, muddy rock debris of mixed sizes, much like a slurry of freshly mixed concrete pouring down a chute.

DECOMPOSITION   The chemical breakdown of rocks and minerals.

DESERT PAVEMENT   An armor of closely fitted stones, one layer thick, on the surface of alluvial material. Basically a residual accumulation of larger fragments owing to removal of fine particles.

DESERT VARNISH   A thin coating rich in iron and manganese on rock surfaces developed by weathering.

DIATOMITE   A sedimentary rock consisting almost entirely of the siliceous skeletons of single-celled algae.

DIKE   A sheet-like body of igneous rock formed by intrusion along a fracture.

DIORITE   A coarse-grained intrusive igneous rock about midway between a granite and a gabbro in chemical and mineralogical composition.

DIP   The direction and degree of inclination (from horizontal) of a sedimentary bed or any other geological planar feature.

DISINTEGRATION   The physical breakup of rocks and minerals.

DOLOMITE   A sedimentary rock composed of the mineral dolomite, a calcium-magnesium carbonate.

DOME   A topographic dome is a roughly circular, upwardly convex land form. A structural dome in sedimentary rocks involves an outward dip or inclination of the beds in all directions. A volcanic dome is a dome-like extrusion of highly viscous lava.

EARTH FLOW   A form of mass movement in which relatively unconsolidated surface material, usually weathered, flows down a hillside.

EMBAYMENT   An indentation along a shoreline, mountain front, or any other natural linear feature.

END MORAINE   A moraine deposited at the lower end of an ice stream or outer end of an ice lobe.

EPICENTER   The spot on the earth's surface directly above the subsurface point at which an earthquake shock originates.

EROSION   The removal of rock material by any natural process.

EXTRUSIVE ROCK   Rock extruded onto the earth's surface, usually in molten condition (lava).

FAN   A deposit, usually alluvial, of rock debris at the foot of a steep slope (mountain face) with an apex at the mountain base (canyon mouth) and a radial, fan-like, divergence therefrom.

FANGLOMERATE The consolidated deposits of an alluvial fan; a variety of conglomerate which is coarse, ill-sorted, and contains angular stones.

FAULT A fracture along which blocks of the earth's crust have slipped past each other.

FAULT RIDGE An elevated, elongate block lying between two essentially parallel faults.

FAULT SLICE A narrow segment of rock caught between two essentially parallel, closely adjacent faults.

FAULT ZONE A zone in the earth's crust consisting of many roughly parallel, overlapping, closely spaced faults and fractures; may be up to several miles wide.

FELDSPAR An abundant rock forming class of minerals composed of aluminum, silicon, oxygen, and one or more of the alkalies, sodium, calcium, and potassium.

FLATIRON RIDGE A linear ridge with one very smooth flank formed by erosion of tilted sedimentary rocks and given a triangular shape by cross cutting canyons.

FLUVIAL Features of erosion or deposition created by running water.

FOLIATION A crude banding formed in rocks by metamorphism, less regular than the bedding of sedimentary rocks.

FORMATION A geological formation is a rock unit of distinctive characteristics which formed over a limited span of time and under some uniformity of conditions. To a geologist it is a rock body of some considerable areal extent which can be recognized, named, and mapped.

GABBRO A dark, coarse-grained intrusive igneous rock richer in iron, magnesium, and calcium and poorer in silica than granite.

GEOPHYSICAL EXPLORATION Subsurface exploration of rocks and structures carried on by indirect means such as gravity or magnetic variations.

GEOTHERMAL Involving heat from within the earth.

GNEISS A coarse-grained metamorphic rock with irregular banding (foliation).

GORGE A narrow, steep-walled passage cut into rock by a stream.

GRABEN A sizeable block of the earth's crust dropped down between two faults steeply inclined inward, giving a keystone shape to the block, longer than it is wide.

GRAIN Used here for a perceptible linear pattern in landscape features of a region, usually reflecting a similar pattern in underlying rock structure.

GRANITE A common, coarse-grained, igneous intrusive rock relatively rich in silica, potassium, and sodium.

GRANITIC A term commonly used for many coarse-grained igneous intrusive rocks not strictly of granite composition.

GRANODIORITE A coarse-grained, igneous intrusive rock half way between a granite and a diorite on the scale of rock composition.

GULLY A small ravine cut by running water.

HANGING VALLEY A tributary valley the floor of which is much higher at its mouth than the floor of the trunk valley.

HOGBACK A ridge composed of a resistant layer within steeply tilted eroded strata.

IGNEOUS ROCKS A class of rocks formed by crystallization from a molten state.

INCLUSION A fragment of older rock inclosed (included) within an igneous rock.

INCOMPETENT A rock which is relatively weak and responds readily to pressure by crumpling or by flow.

INTERMITTENT STREAM One which does not have a continuous or perennial flow.

INTRUSIVE Rocks or rock masses which have been intruded or injected into other rock, usually in a molten state.

LATERAL FAULT One on which the displacement is sidewise rather than up-down.

LATERAL MORAINE A ridge-like deposit of bouldery ill-sorted debris laid down along the lateral margin of a valley glacier.

LAVA The term is used both for molten rock material extruded onto the earth's surface and for the consolidated (crystallized) rock.

LEFT LATERAL FAULT One on which the opposing block appears to have moved to the left, no matter which side you stand on.

LIMB One of the two sides of an anticline or syncline.

LIMESTONE A sedimentary rock composed wholly or almost wholly of the mineral calcite.

MAGMA Molten rock within the earth's crust.

MARBLE Recrystallized limestone or dolomite; a metamorphic rock.

MARINE The ocean environment; marine sediments are those deposited in the ocean.

MASS MOVEMENTS The movement, usually down slope, of a mass of rock or rock debris by gravity, not transported by some other agent such as ice or water.

MATRIX The fine-grained constituents of a rock in which coarser particles are embedded.

MESA A flat-topped tableland with steep sides.

MESOZOIC One of the eras of the geological time scale (*see* Appendix A) extending from 70 to 230 m.y. ago.

METAMORPHIC ROCKS Those which have undergone such marked physical change because of heat or pressure or both as to be distinct from the original rock. The process is *metamorphism*.

METAVOLCANIC Rocks formed by metamorphism of volcanic materials.

M.Y. An abbreviation for a million years.

MINERAL A homogeneous, naturally occurring, solid substance of inorganic composition, consistent physical properties, and specified chemical composition.

MONOLITHOLOGIC BRECCIA A breccia formed of fragments of only one kind of rock.

MONOMINERALLIC ROCK One composed of only one mineral; for example, limestone and dolomite.

MORAINE A deposit of coarse, ill-sorted rock debris laid down by glacial ice without intervention of any other agent, such as running water.

MUDFLOW A form of mass movement involving the flow of mud, usually containing coarser rock debris, in which instance the term debris flow is equally applicable.

MUD POT A shallow, hot-spring pit filled with bubbling mud.

MUDSTONE A fine-grained sedimentary rock which is hard to characterize as shale or siltstone because of massiveness or poor sorting.

NOSE *See* anticlinal nose.

OBLIQUE AIR PHOTO One taken with the axis of the camera tilted from vertical. If the horizon shows, it is a high-oblique photo.

OBSIDIAN Natural volcanic glass. Lava which cooled so rapidly that it didn't crystallize.

ODOMETER An instrument for measuring distance.

ORE DEPOSIT An accumulation of metallic minerals that can be mined at a profit. The minerals are termed *ore minerals,* and the aggregate is termed *ore.*

OUTCROP An exposure of bedrock at the surface.

PALEOZOIC A major era of the geological time scale embracing the interval from 230 to 600 m.y. (*see* Appendix A).

PENDANT A large mass of metamorphic rock within a younger intrusive rock, thought to have hung down into the original intrusive body from the roof of the intrusive chamber.

PEDIMENT A relatively smooth, gently sloping surface produced by erosion at the foot of a steeper face, usually a mountain.

PEGMATITE A very coarse-grained igneous rock formed by the fluids given off in the late stage of crystallization of an igneous body; most often close to granite in composition.

PLACER DEPOSIT A water laid accumulation of rock debris containing a concentration of heavy, physically and chemically resistant, valuable mineral such as diamond, gold, or platinum. Such minerals are described as *placer minerals.*

PLAYA The flat, smooth floor of a dry lake in desert regions.

PLEISTOCENE An epoch within the Cenozoic Era of the geological time scale, (*see* Appendix A). Usually taken to embrace the last 2 million years.

PLUG A small, cylindrical, near-surface, igneous intrusive body.

PLUNGE The inclination from horizontal of the long axis of a fold or warp.

PLUVIAL PERIOD An interval of cooler, wetter conditions in a dry region, coincident with a phase of glaciation in colder, better-watered areas.

POTASSIUM-ARGON A method of absolute dating of rocks and minerals using the ratio of radioactive potassium to its daughter product, the argon 40 isotope.

POTHOLE A narrow cylindrical hole worn into solid rock by a fixed vortex in a stream.

PRECAMBRIAN All rocks older than Paleozoic (*see* Appendix A).

PUMICE Frothy rock glass, so light that it floats.

PYROCLASTIC Hot or firey (pyro) fragmental (clastic) debris thrown out of an explosive volcanic vent.

PYROXENE A common igneous- and metamorphic-rock family of minerals, often green to black, and ranging widely in composition.

QUARTZ One of our most common minerals, hard and chemically resistant, composed of silicon and oxygen (SiO_2).

QUARTZITE A rock formed by metamorphism of sandstone, which is hard, coherent, and consists of quartz.

RADIOACTIVE The property of some elements to spontaneously change into other elements with the emission of charged particles, usually accompanied by generation of heat.

RADIO-CARBON The radioactive isotope of carbon (^{14}C) which disintegrates at a known rate. It is used to determine geological ages up to about 40,000 years.

RARE EARTHS The oxide compounds of rare-earth elements, such as cerium, ytterbium, neodymium, and others.

RECHARGE WELL A well designed for injection of fluids into the ground.

RELIEF Topographic relief is the difference in elevation of contiguous parts of a landscape, valley to peak.

RHYOLITE An extrusive igneous rock of granitic composition, fine-grained, often light-colored to red.

RIFT As used here, refers to the shallow topographic trench, a mile or two wide, along the trace of a major fault.

RIGHT-LATERAL FAULT One on which the displacement of the opposing block appears to have been to the right, no matter on which side the observer stands.

RILLENSTEINE A stone with small, interlacing, worm-like solution channels on its surface. They form on soluble rocks, most commonly limestone.

ROCK An aggregate of minerals.

ROCK CLEAVAGE The facility to break along parallel smooth planes within a mass of rock.

ROCKFALL The relatively free fall of rock masses from steep bedrock faces.

ROCK GLACIER An accumulation of large angular blocks of rock, usually lobate in form with steep margins, that moves slowly by creep.

SAGPOND A pond occupying a depression along the trace of a major fault, usually where a block within the zone has sunk.

SANDSTONE A sedimentary rock formed by cementation of sand-size particles.

SCARP A straight steep bank or face which can be a few feet to thousands of feet high, like the east face of the Sierra Nevada.

SCHIST A finer-grained and more thinly and regularly foliated metamorphic rock than gneiss.

SCORIA Small fragments of porous volcanic rock thrown out of an explosive volcanic vent. Usually black or red and up to $1\frac{1}{2}$ inches in diameter.

SEDIMENTARY ROCKS A class of rocks of secondary origin, made up of transported and deposited rock and mineral particles and of chemical substances derived from weathering.

SEPTUM An older mass of metamorphic rock separating two adjacent intrusive igneous bodies.

SERPENTINITE A rock consisting largely of the mineral serpentine, a hydrous magnesium silicate, produced by alteration of igneous rocks rich in iron and magnesium.

SHALE A sedimentary rock consisting largely of very fine mineral particles, laid down in thin layers.

SILICEOUS Rich in silica, SiO_2.

SILTSTONE A fine-grained, well-bedded sedimentary rock composed of silt, finer than sand and coarser than clay.

SLATE A weakly metamorphosed rock derived from shale by compaction with the development of closely spaced, smooth, parallel breaking surfaces (slaty cleavage).

SOAPSTONE A massive, soft, slippery rock composed of the mineral talc, a hydrous magnesium silicate.

SORTING The arrangement of particles by size.

SPUR The subordinate ridges extending from the crest of a larger ridge.

STRATA Layers of a sedimentary rock. Bedded rocks are *stratified*.

STREAM CAPTURE The diversion of the

headwaters of a stream owing to headward growth of an adjacent stream.

STRUCTURE Phenomena that determine the geometrical relationships of rock units, such as folds, faults, and fractures.

SUPERIMPOSED STREAM One which has cut down through an overlying mantle into rocks of different character and structure.

SYENITE An intrusive igneous rock much like granite but lacking or very low in quartz.

SYNCLINE A down-fold in layered rocks which is concave upward. Beds on the flanks are inclined inward.

TERRACE A geometrical form consisting of a flat tread and a steep riser or cliff. Stream terraces, lake terraces, marine terraces, and structural terraces are distinguished in geology.

TERRESTRIAL Deposits laid down on land as contrasted to the sea; terrestrial conditions as compared to marine conditions.

TERTIARY A period of the Cenozoic Era (see Appendix A) embracing the time from 70 to 2 m.y. ago.

THRUST FAULT A gently inclined fault along which one block is thrust over another.

THRUST PLATE The upper block of a thrust fault.

TILL Ill-sorted, mixed fine and coarse rock debris deposited directly from glacial ice.

TRAVERTINE An accumulation of calcium carbonate formed by deposition from ground or surface waters, commonly porous and cellular.

UNCONFORMITY A surface of erosion separating younger strata from older rocks.

VARNISH See desert varnish.

VEIN A sheet-like deposit of mineral matter along a fracture.

VENTIFACT A stone whose shape and surface characteristics have been modified by natural sandblasting.

VERTICAL AIR PHOTO One taken with the axis of the camera pointed straight down toward the ground.

VOLCANIC ASH Fine-grained (less than ⅛ inch diameter) volcanic debris, often glassy, explosively erupted from a volcanic vent.

VOLCANIC CINDERS Like volcanic ash but coarser, ⅛ to 1 inch. Fragments are highly porous.

VOLCANIC TUFF A compacted deposit consisting of ash, cinders, and occasionally larger fragments of solid volcanic rock. If the latter are numerous, it is known as a tuff-breccia.

WARP A part of the earth's crust which has been broadly bent.

WATER GAP A gap in a ridge, cut and still occupied by the stream that cut it.

WATER TABLE The level beneath the ground surface below which all openings in rocks are filled with water.

WIND GAP A gap or saddle in a ridge now abandoned by the stream that cut it.

WINEGLASS CANYON A canyon cut into the steep face of a mountain range. The fan at the mountain foot is the base, the gorge approaching the canyon mouth is the stem, and the headwaters basin is the bowl.

ZIRCON A mineral found in small amounts in many igneous and metamorphic rock, a zirconium silicate and a gemstone. Chemically and mechanically tough.

APPENDIX C
ANNOTATED BIBLIOGRAPHY

General Background

BLOOM, ARTHUR L. *The Surface of the Earth.* Englewood Cliffs, N. J.: Prentice-Hall, Inc., 1969.

A simplified, up-to-date treatment of processes shaping features of the earth's surface (paperback).

FENTON, C. A., and M. A. *The Rock Book.* New York: Doubleday, Doran and Co., 1940.

Although old, this highly readable book presents material on rocks and minerals in a style attractive to nonprofessionals (hardcover).

PEARL, RICHARD M. *Geology.* New York: Barnes and Noble, Inc., 1966.

An introduction to both physical and historical geology presented in an easily understandable style, a brief treatment of the essential basic elements (paperback).

POHOPIEN, K. M. *An Introduction to the Megascopic Study and Determination of Minerals and Rocks.* Dubuque, Iowa: Wm. C. Brown Company Publishers, 1969.

For people interested in rocks and minerals this field guide is a useful aid. It provides determinative tables and illustrations helpful in mineral and rock identification (paperback).

SHELTON, JOHN S. *Geology Illustrated.* San Francisco: W. H. Freeman and Co., 1966.

A superbly illustrated book dealing with a wide variety of geological features and phenomena, many in southern California (hardcover).

TURNER, DANIEL S. *Applied Earth Science.* Dubuque, Iowa: Wm. C. Brown Company Publishers, 1969.

In this day of concern about the relevance of science and the protection of our environment this artfully written booklet should have wide appeal (paperback).

TUTTLE, SHERWOOD O. *Landforms and Landscapes.* Dubuque, Iowa: Wm. C. Brown Company Publishers, 1970.

A condensed but comprehensive presentation relating landforms and landscape features to genetic processes (paperback).

Southern California (General)

BAILEY, EDGAR H., ed. *Geology of Northern California,* Bulletin 190. California Division of Mines and Geology, 1966.

A companion volume to the southern California book. Chapters 5 (Great Valley), 6 (Coast Ranges), and especially 4 (Sierra Nevada) are of particular interest in relation to areas described in our booklet (hardcover).

DOWNS, THEODORE. *Fossil Vertebrates of Southern California.* Berkeley: University of California Press, 1968.

A small pocketbook in popular style for nonprofessional readers, some striking color illustrations (paperback).

GEOLOGIC MAP OF CALIFORNIA. California Division of Mines and Geology.

This map consists of twenty-six separate sheets on a scale of 4 miles to the inch. The sheets can be separately purchased from the California Division of Mines and Geology, Ferry Building, San Francisco, California 94111, for $1.50 plus tax. The sheets of primary use in the areas treated herein are: Mariposa, Fresno, Death Valley, Bakersfield, Trona, Kingman, Los Angeles, San Bernardino, Needles, Santa Ana, and Salton Sea. This is the finest state geological map extant.

GEOLOGIC MAP OF CALIFORNIA. U. S. Geological Survey Miscellaneous Geological Investigations Map I-512.

171

This excellent, small-scale map showing natural provinces, faults and rock distribution can be obtained for only 25 cents by an order addressed to: U. S. Geological Survey, Distribution Section, Federal Center, Denver, Colorado 80255, or purchased at 7638 Federal Building, 300 N. Los Angeles St., Los Angeles.

HINDS, N. E. A. *Evolution of California Landscape,* Bulletin 158. California Division of Mines and Geology, 1952.

Provides description of features within the various natural provinces, lavishly illustrated with photographs (hardcover).

JAHNS, R. H., ed. *Geology of Southern California,* Bulletin 170. California Division of Mines and Geology, 1954.

This is the definitive professional work on southern California. It is a massive compilation of articles, maps, and guidebooks dealing with almost every conceivable aspect of the region's geology (encased).

LACOPI, ROBERT. *Earthquake Country.* Menlo Park, Calif.: Lane Book Co., 1964.

A reliable description of California faults and earthquakes prepared for the layman, with good format and excellent illustrations (hardcover).

OAKESHOTT, GORDON B. *California's Changing Landscapes.* New York: McGraw-Hill Co., 1971.

A beautifully illustrated presentation of California geology written by a man who has devoted a lifetime to the subject. In addition to descriptions of provinces, areas, and features, it provides solid elementary geological background at a somewhat more professional level than attempted herein (available in hardcover or paperback).

Russ Leadabrand Guidebooks. Los Angeles: Ward Richie Press.

An inexpensive series of small pocketbooks describing aspects of the natural and human history of various parts of southern California by a man who knows the country like the palm of his hand and loves it deeply. The following are particularly pertinent to areas or features treated herein:

A Guidebook to — The San Gabriel Mountains
The San Bernardino Mountains
The Sunset (Peninsular) Ranges
The Mojave Desert (including Death Valley)
The Southern Sierra Nevada

Exploring California Byways:
I — Kings Canyon to Mexican Border
III — Desert Country

Specific Areas

Los Angeles Basin

WOODRING, W. P.; BRAMLETTE, M. N.; and KEW, W. S. W. *Geology and Paleontology of Palos Verdes Hills,* Professional Paper 207. U. S. Geological Survey, 1946.

The recognized standard reference on Palos Verdes Hill by three extremely competent men.

YERKES, R. F. and McCULLOH, T. H., SCHOELLHAMER, J. E., and VEDDER, J. G. *Geology of the Los Angeles Basin, California,* Professional Paper 420-A. U. S. Geological Survey, 1965.

The most authoritative introduction to the geology of this area ever prepared.

Death Valley

HUNT, CHARLES B., and MABEY, DON R. *Stratigraphy and Structure, Death Valley, California,* Professional Paper 494-A. U. S. Geological Survey, 1966.

The most modern, thorough, and competent professional geological publication available on the valley and its immediate environs.

KIRK, RUTH. *Exploring Death Valley.* Stanford, Calif.: Stanford University Press, 1965.

A small guidebook on the human and natural history of Death Valley with much useful information for visitors and travelers, written by a lady who knows the country well (paperback).

MAXSON, JOHN H. *Death Valley, Origin and Scenery.* Bishop, Calif.: Chalfant Press, 1963.

Death Valley as seen in terms of its geological evolution, a general treatment more than a field trip guide, many excellent photographs (paperback).

Owens Valley—Sierra Nevada

BATEMAN, P. C., and WAHRHAFTIG, CLYDE. "Sierra Nevada Province." Chapter IV in *Geology of Northern California,* Bulletin 190. Division of Mines and Geology, 1966.

This is the most up-to-date authoritative statement available concerning the geology of the Sierra Nevada.

HILL, DAVID W. *Inyo Skyline,* Bishop, Calif.: Chalfant Press, 1965.

A road-log guide to topographic and other features seen on a highway trip through Owens Valley (paperback).

————. *Mono Skyline.* Bishop, Calif.: Chalfant Press, 1969.

A road-log guide to topographic and other features north from Bishop (paperback).

RINEHART, C. D., and ROSS, D. C. *Geology and Mineral Deposits of the Mount Morrison Quadrangle, Sierra Nevada, California.* Professional Paper 385. U. S. Geological Survey, 1964.

This publication is cited as an example of the many excellent professional papers and bulletins issued by the U. S. Geological Survey. The area is adjacent to the U. S. Highway 395 route to Mammoth.

SCHUMACHER, GENNY, ed. *Deepest Valley, A Guide to Owens Valley, Its Lakes, Roadsides, and Trails.* Berkeley: Wilderness Press, 1969.

An authoritative, well written, and nicely illustrated guide to the natural and human history of Owens Valley and environs (paperback).

————, ed. *The Mammoth Lakes Sierra, A Handbook for Roadside and Trail.* Berkeley: Wilderness Press, 1969.

Equal in calibre to the booklet on Owens Valley and fulfilling a similar function for the Mammoth Lakes region (paperback).

INDEX
LOCALITIES AND FEATURES

175